Willi Schönauer
Wolfgang Gentzsch (Eds.)

The Efficient Use
of Vector Computers
with Emphasis on
Computational Fluid Dynamics

W0106295

Notes on Numerical Fluid Mechanics
Volume 12

Series Editors: Ernst Heinrich Hirschel, München
Maurizio Pandolfi, Torino
Arthur Rizzi, Stockholm
Bernard Roux, Marseille

Manuscripts should have well over 100 pages. As they will be reproduced photomechanically they should be typed with utmost care on special stationary which will be supplied on request. In print, the size will be reduced linearly to approximately 75 %. Figures and diagrams should be lettered accordingly so as to produce letters not smaller than 2 mm in print. The same is valid for handwritten formulae. Manuscripts (in English) or proposals should be sent to the general editor Prof. Dr. E. H. Hirschel, MBB-LKE 122, Postfach 80 11 60, D-8000 München 80.

Willi Schönauer
Wolfgang Gentzsch (Eds.)

The Efficient Use
of Vector Computers
with Emphasis on
Computational
Fluid Dynamics

A GAMM-Workshop

Springer Fachmedien Wiesbaden GmbH

Produced by Lengericher Handelsdruckerei, Lengerich

ISBN 978-3-528-08086-0   ISBN 978-3-663-13912-6 (eBook)
DOI 10.1007/978-3-663-13912-6

PREFACE

The GAMM Committee for Numerical Methods in Fluid Mechanics organizes workshops which should bring together experts of a narrow field of computational fluid dynamics (CFD) to exchange ideas and experiences in order to speed-up the development in this field. In this sense it was suggested that a workshop should treat the solution of CFD problems on vector computers. Thus we organized a workshop with the title "The efficient use of vector computers with emphasis on computational fluid dynamics". The workshop took place at the Computing Centre of the University of Karlsruhe, March 13-15,1985. The participation had been restricted to 22 people of 7 countries. 18 papers have been presented.

In the announcement of the workshop we wrote: "Fluid mechanics has actively stimulated the development of superfast vector computers like the CRAY's or CYBER 205. Now these computers on their turn stimulate the development of new algorithms which result in a high degree of vectorization (scalar/vectorized execution-time). But with 3-D problems we quickly reach the limit of present vector computers. If we want e.g. to solve a system of 6 partial differential equations (e.g. for u, v, w, p, k, ε or for the vectors u, curl u) on a 50x50x50 grid we have 750.000 unknowns and for a 4th order difference method we have circa 60 million nonzero coefficients in the highly sparse matrix. This characterizes the type of problems which we want to discuss in the workshop".

When the first CRAY-1 (12,5 nsec cycle time) was delivered in 1976, it was an "exotic", a "supercomputer". It seemed that only a very few large research establishments would be able to afford such a computer. Now, roughly 9 years later, the CRAY-1 is history, it has been replaced by the CRAY X-MP (9,5 nsec cycle time) and the CRAY-2 is now on the market (4 nsec cycle time). But the most unexpected thing in this development is, that circa 70 CRAY-1's have been delivered. From the other "early" vector computer which reached commercial significance, the CYBER 205 of CDC (20 nsec cycle time), circa 30 units have been delivered. In Germany in the field of research there are presently installed 4 vector computers at universities and 3 vector computers in research establishments. In the USA a program has been initiated to give access to all universities to vector computing centres. In the commercial sector all large industrial companies in the field of aircraft, automobile or oil will have one or several vector computers if they do not have already.

Ultimately the extreme computational speed of the vector computers results from the parallelism which is inherent in most largescale computations (mostly this is some form of matrix manipulation). But the programmer has to write his programs in such a form that the vector computer is able to transform the language instructions into vector operations, otherwise it will compute only in scalar mode, i.e. like a usual general purpose computer. But we are still on the way to learn this "parallel computing". In this sense the goal of the workshop was to put together the experiences of the participants that they might present their own experiences to the other participants and, on the other hand, learn themselves from the experiences of the others.

We have mentioned above that a 3-D grid of 50x50x50 is the limit for the present generation of vector computers. But for the description of the flow around an automobile or a whole aircroft in sufficient details a grid of 500x500x500 would be needed. This requires the 1000-fold computational speed of the present generation of vector computers. If we denote the first generation by the "100 MFLOPSgeneration" (MFLOPS=million floating point operations per second), we have now by the recently announced CRAY-2 the 1 GFLOPS (GFLOPS= 1000 MFLOPS), around 1988 10 GFLOPS and perhaps 1992 the 100 GFLOPS. This will be possible only by a new type of parallelism: the multiprocessor vector computers. But these speeds can be achieved only if the main memories are large enough or if there are fast and large enough secondary storages to store the immense data sets of such large problems. The development of superfast computers will surely never end. Neil Lincoln who has designed the CYBER 205 and who is now designing the successor, the ETA[10], gave the following definition of a supercomputer: "A supercomputer is a computer which is only one generation behind the requirements of the large scale users". These remarks may conclude the definition of the background of the workshop.

The papers of the workshop are not presented in these proceedings as usual in alphabetic order of the authors. We had composed the program of the workshop by grouping the papers according to the subjects which were treated, to the vector computers and type of solution methods which were applied. And in the same sequence the papers are presented in these proceedings. We decided also that the participants should write their papers immediately after the workshop and that they should include suggestions which were obtained from the discussions. In this way, most recent results have been included.

In an introductory paper the hard- and software of the most relevant vector computers with their typical properties and deficiencies are presented. Then a group of 6 papers deals with the full potential and Euler equations. It is very informative to see how these problems are treated on the different types of vector computers by different types of discretization methods. Then follows a paper which treats the shallow water equations. The solution of the 2-D and 3-D Navier Stoker equations is presented in a group of 6 papers. Then follow two papers of the field of meteorology and climate modelling. A further paper reports on experiences with the testing of different linear algebra algorithms on vector computers in the frame of CFD problems.

In the final paper which was ultimately worked out during the workshop a summary and a conclusion of the workshop are presented. This paper tries to classify the problems which were treated on the different vector computers and the methods which were used, states the problems and points out the direction which further developments should take. We hope that these proceedings of the workshop will enable the reader to profit in the same way as we did from the cumulated experience which was gathered at the workshop.

June 1985   Willi Schönauer   Wolfgang Gentzsch

CONTENTS

INTRODUCTION TO THE WORKSHOP:

SOME BOTTLENECKS AND DEFICIENCIES OF

EXISTING VECTOR COMPUTERS AND THEIR

CONSEQUENCES FOR THE DEVELOPMENT OF

GENERAL PDE SOFTWARE

W.Schönauer, E.Schnepf

Rechenzentrum der Universität Karlsruhe

Postfach 6380, D-7500 Karlsruhe 1, West-Germany

SUMMARY

     In the first part of this introductory paper a review of
the different types of vector computer programs and algorithms
is presented. Then the hardware of the presently most relevant
vector computers together with their bottlenecks will be discussed as well as the trends in development. The weakest point
of the existing vector computers is the compiler, this problem
is closely related to the lacking vector statements of FORTRAN
77. Therefore some proposals of FORTRAN 8X will be presented.
In the second part of this paper the authors report about some
experiences which have been obtained in the development of
"black-box software" for PDE's. There will be given a pragmatic definition of a "data flow algorithm", and the separation
of data selection and processing will be demonstrated for the
evaluation of difference formulae. Then the i/o bottleneck is
discussed for the iterative solution of large linear systems
in diagonal storing. There will be presented our view of portability of software for different vector computers. The numerical example for ILU-preconditioning demonstrates how vectorisation might invert experiences gained on general purpose
computers. Finally some examples demonstrate the application
of the FIDISOL program package on different vector computers.

1. HISTORICAL DEVELOPMENT OF PROGRAMS

AND ALGORITHMS

     If we consider in a sort of historical review the develop-
ment of user programs on vector computers, we might distinguish
between 3 generations of vector computer software:

1

1.1 Unchanged general purpose computer programs. These programs have been developed on general purpose computers for
general purpose computers. They have many nonvectorisable
loops and a bad data structure. On present vector computers
they obtain (eventually) 10-15 MFLOPS (million floating-
point operations per second).
1.2 "Hand vectorised" general purpose computer programs. The
user tries to remove, mostly on the level of the coding,
the reasons for the nonvectorisation of the loops, using
the information of the compiler who tells to the user which
loops why have not been vectorised. But usually there remains still the bad data structure. Such programs may obtain 15-20 MFLOPS.
1.3 Special vector computer programs. These programs have been
designed from scratch for vector computers, with an optimal data structure and with optimal algorithms. Such programs may obtain a sustained rate of 40-80 MFLOPS.

In a similar way we might distinguish between three generations of vector computer algorithms:
2.1 Unchanged general purpose computer algorithms, with many
recursions, with short vector length, not designed for vectorisation.
2.2 "Vectorised" algorithms, still incore, only designed to
solve the old 2-D or explicit 3-D problems very fast, but
they are not suited for large data sets.
2.3 Special vector computer algorithms, designed for large 3-D
data sets, implicit solution methods, optimal data structure also for out-of-core. The performance of such algorithms
is not limited by the speed of the vector pipes but rather
by the i/o bottleneck of most installed vector computers.

The conclusions which we can draw from this simplifying
historical review of user programs and algorithms are the following ones:

• "Old" general purpose computer software wastes the power of
a vector computer. This danger holds above all for large industrial program packages into which have been invested many
many man years of programming. People who judge the power of
a vector computer from the generations 1.1 or 1.2 mentioned
above, may be well disappointed. It might be interesting to
note that in comparison to a CYBER 205 a CRAY-1 usually will
deliver shorter execution times for such programs because the
CRAY-1 is not so "sensitive" to "bad" software.

• The power of a vector computer grows considerably with the
skill of the user, much more than would be possible on a general purpose computer. Going from 10 MFLOPS for "old" software to 80 MFLOPS for special vector computer software means
an increase of the huge power of a vector computer by a factor of 8! It is interesting to note that for special vector
computer software of the generation 1.3 the CYBER 205 usually will win over the CRAY-1, above all if the algorithms are
designed for long vectors.

• The increased speed of special vector computer software immediately allows to attack <u>larger</u> problems, but these problems then produce larger data sets.

- Large 3-D problems (and the real world is 3-D!) are on the present generation of installed vector computers more severly limited by the i/o than by the arithmetic speed of the vector pipes because the size of the main storage is usually too small to allow an incore treatment of 3-D problems. This misbalancing results from the extremely expensive technology which was used in the main memories.

- An efficient use of a vector computer is guaranteed only by "data flow" type algorithms which maintain a continuous data flow through the pipes.

- Therefore we need for vector computers well trained users, who must be much better trained than for general purpose computers.

- It might be useful at this point to mention the literature which can help to train our users better. There is the excellent book of Hockney/Jesshope [1] which is as a general introduction to vector computers up to now the best reference. The book of Gentzsch [2] is a natural extension of the introductory literature as it contains many useful applications and is in some sense similar to the intentions of this workshop.

2. THE HARDWARE OF THE MOST RELEVANT

VECTOR COMPUTERS

2.1 CRAY-1

The CRAY-1 has a cycle time of 12.5 nsec (nano sec = 10^{-9} sec), it has $n_{1/2} \approx 20$. Hockney's $n_{1/2}$ is the vector length which is needed to get half the real peak performance [1]. The CRAY-1 is now history, it is no longer produced. Between 60 and 70 units have been delivered. The CRAY-1 is a milestone in the development of vector computing. It started as an "exotic" computer and then succeeded in making the breakthrough for commercial vector computing. From all the vector computers mentioned in this paper the value $n_{1/2} \approx 20$ is the smallest one. This has the consequence that programs need not be designed for long vectors. Thus also programs of the generations 1.1 and 1.2 of section 1 run "efficiently" on the CRAY-1.

The CRAY-1 is the prototype of a <u>register-to-register</u> machine: The functional units (pipes) operate only with operands which are stored in one of the 8 vector registers (each V-register has 64 elements). This is the great chance and great danger of this type of computer at the same time: The efficiency depends largely on the <u>chaining</u>, i.e. the overlapping of

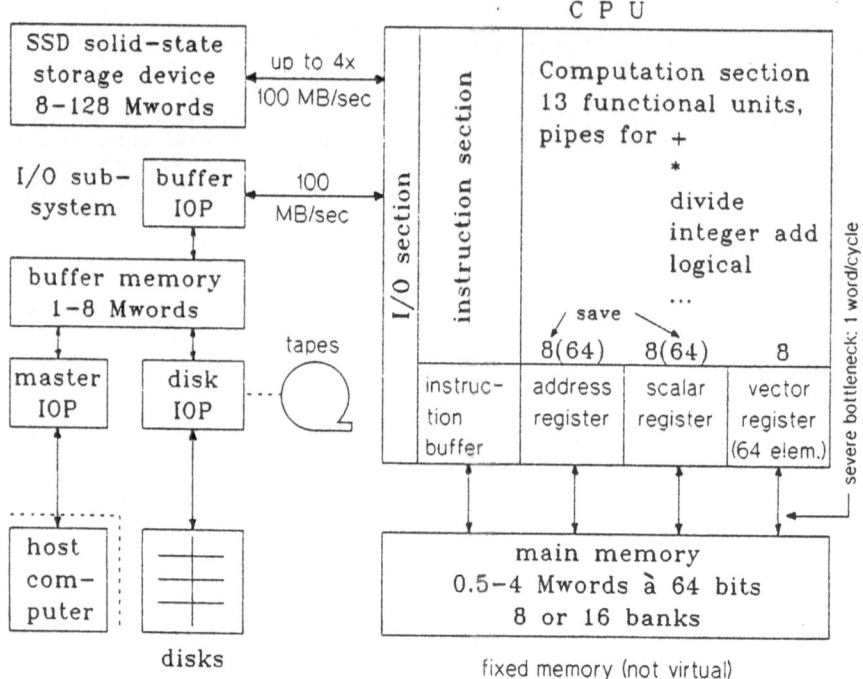

Fig. 2.1 Schematic view of the CRAY-1 S.

operations and register loading. If all operands are in the vector registers, the operations are "uncoupled" from the main memory. But if the operands are not present, the pipes have to wait. It is obviously still a severe problem for the compiler to do an efficient chaining.

The CRAY-1 has a severe bottleneck: There is only <u>one</u> word transfer/cycle between main memory and V-registers to <u>or</u> from the main memory. But a simple operation like

$$c_i = a_i + b_i \qquad\qquad (2.1)$$

needs 3 transfers/cycle, two loads and one store, but there is possible only one transfer/cycle. Therefore for such simple operations the peak performance is not limited by the 80 MFLOPS of the arithmetic pipe but by $80/3 \approx 27$ "MFLOPS" of the data path between main memory and V-registers. This bottleneck has also the consequence that long vectors are subdivided by the hardware into sections of 64 elements and after each section of 64 elements there is a restart of the pipe, leading to a saw-tooth curve for the MFLOPS-rate. It is possible to process noncontiguous vectors with constant stride, but this bears the danger of bank conflicts. The CRAY-1 S has a bipolar memory with 4 cycles bank-busy-time and the CRAY-1 M has a MOS me-

mory with 8 cycles bank-busy-time. If a bank is addressed before its bank-busy-time is over there are lost cycles for waiting. Thus also on the CRAY-1 the use of contiguous vectors
should be recommended to avoid bank conflicts. Without the fast
secondary storage SSD there is a severe i/o bottleneck, i.e.
the data path between main memory and secondary storage is too
narrow compared to the speed of the vector pipes, leading to
problems for out-of-core programs. There are no hardware instructions for gather/scatter which are needed for all indirect
addressing problems.

2.2 CRAY X-MP/2

The denotation /2 means two processors. The cycle time of
the CRAY X-MP is 9.5 nsec, it has a value $n_{1/2} \simeq 50$ which is
already 2.5 times that of the CRAY-1. The bottleneck

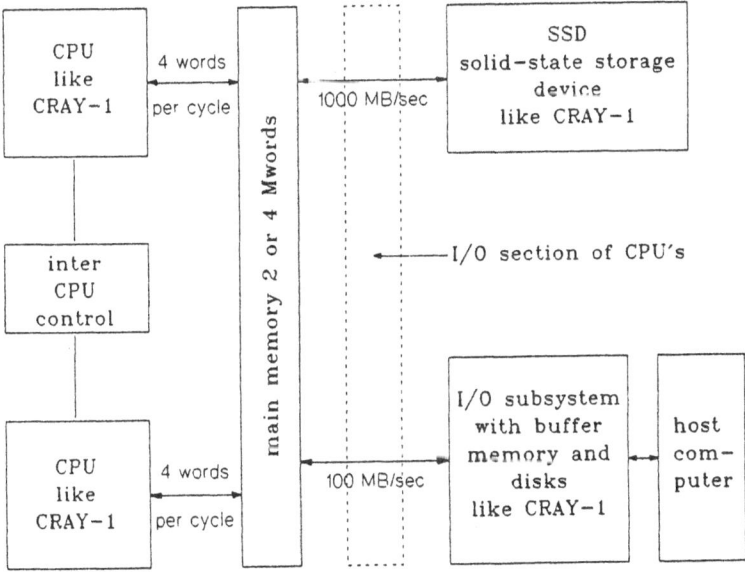

Fig. 2.2 Schematic view of the CRAY X-MP/2. The SSD and the i/o
        subsystem are in reality connected to the i/o-section
        of the CPU's.

between main memory and V-registers has been removed: There are
for each processor 3 transfers/cycle between main memory and
V-registers and one transfer/cycle for i/o. Although the vector
length of each V-register is 64 elements there might be a continuous data flow from the main memory through the V-register
to the pipe and back to another V-register and from there to
main memory that there is not a restart after 64 elements as
in the CRAY-1. Although theoretically the bottleneck between
main memory and V-registers has been removed one has to be careful. If we consider a X-MP/22, i.e. with two processors and

5

two Mwords (2 million words) main memory, the main memory has 16 banks and because of the 4 cycles bank busy time there are possible only 16/4 = 4 transfers/cycle. If we assume that each processor executes the simple operation (2.1) there are needed 6 transfers/cycle. But because there are possible only 4 transfers/cycle the peak rate would not be determined by the 105.3 MFLOPS of the pipes but by the $105.3 \cdot 6/4 \simeq 70$ "MFLOPS" of the main memory modularity. Only the X-MP/24 with 4 Mwords and 32 banks has enough modularity of the main memory.

The CRAY X-MP/1 has only one processor and a MOS main memory with 8 cycles bank-busy-time. It has now replaced the CRAY-1.

The CRAY X-MP/4 has 4 CPU's and 8 Mwords main memory. It can be equipped with two 1000 MB/sec channels to the SSD. The CPU's of the X-MP/4 (only!) have now also hardware instructions for gather/scatter and compressed index instructions. This means that this computer will behave much more efficient e.g. for finite element calculations which needed in an extensive manner indirect addressing.

With the X-MP/2 and X-MP/4 CRAY makes a first step into the direction of a MIMD (multiple instruction stream / multiple data stream) computer. There are special registers for inter CPU control to exchange synchronisation information. CRAY offers software that several processors can work for one single job, i.e. for multiprocessing, which they call "multitasking".

2.3 CYBER 205

The CYBER 205 of CDC (Control Data Corpor.) has a cycle time of 20 nsec. The computer can be purchased with 1 or 2 or 4 vector pipes and the corresponding values of $n_{1/2}$ are 50 or 100 or 200. This means that for a 4-pipe CYBER 205 the vector length to get half the peak performance is 10 times as large than for the CRAY-1, i.e. such a computer can be efficiently used only with long vectors. A bit of history: Another mile stone in the development of vector computers was the STAR 100 which taught us that a vector computer without a fast scalar unit is not efficient. Equipped with a fast scalar unit it became the CYBER 203 and again equipped with faster vector pipes it became the CYBER 205.

The CYBER 205 is the prototype of a memory-to-memory machine: The vector pipes operate on data from the main memory to the main memory, the stream unit acting as an operand buffer. Vectors must be contiguous storage locations in the main memory. The main memory has the huge virtual address space of $2 \cdot 10^{12}$ words. But the computer has a severe drawback: The maximal (hardware) vector length is 64K-1 = 65535 (what a pity!) which is even not hidden by the compiler. The reason is the 16 bits for the vector length register. The computer has separate pipes for scalar and vector arithmetics which can operate in parallel for independent operations. A great advan-

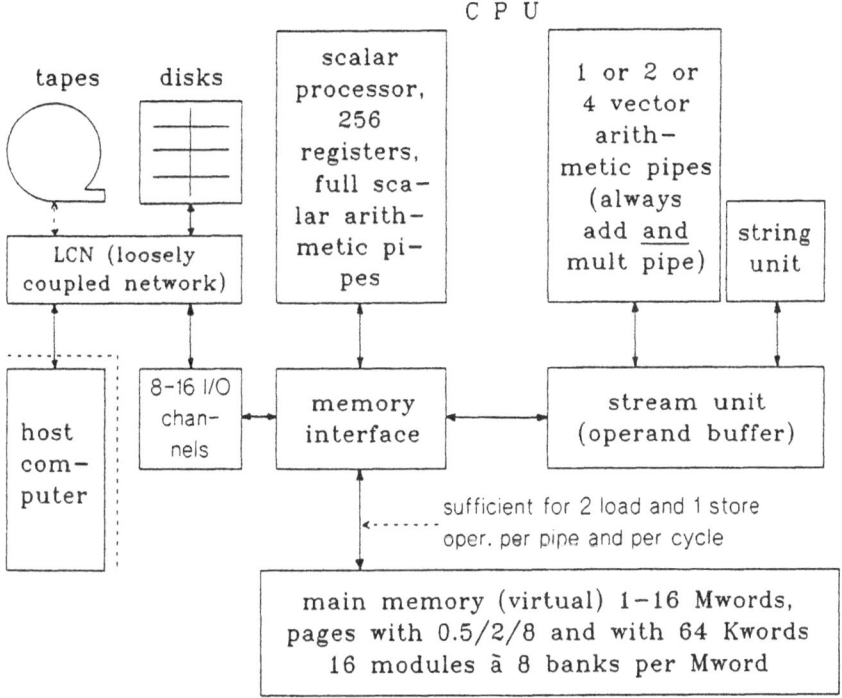

Fig. 2.3 Schematic view of the CYBER 205.

tage of the CYBER 205 are the sophisticated hardware instructions for gather/scatter, compress/expand/merge which allow for a relatively efficient indirect addressing. There is no fast secondary storage available, thus there is a severe i/o bottleneck. The CYBER 205 has the possibility of 32-bit arithmetic with twice the speed of the 64-bit operations.

2.4 FUJITSU VP 100 and VP 200

The VP 200 has "internally doubled" pipes, i.e. parallel pipes compared to the VP 100. The cycle time is 15 nsec for the scalar unit and 7.5 nsec for the vector unit. The value of $n_{1/2}$ is not yet available, but because of the parallel pipes $n_{1/2,VP\ 200} = 2 \cdot n_{1/2,VP\ 100}$. The Japanese have developed the basic technology in a common research project between the hard competitors Fujitsu, Hitachi and NEC and thus have saved a considerable amount of development costs. They have combined the advantages of the CRAY and CYBER vector computers. We consider this a "fruitful" competition which will force the "Americans" to develop better and faster vector computers and thus speeds up the development.

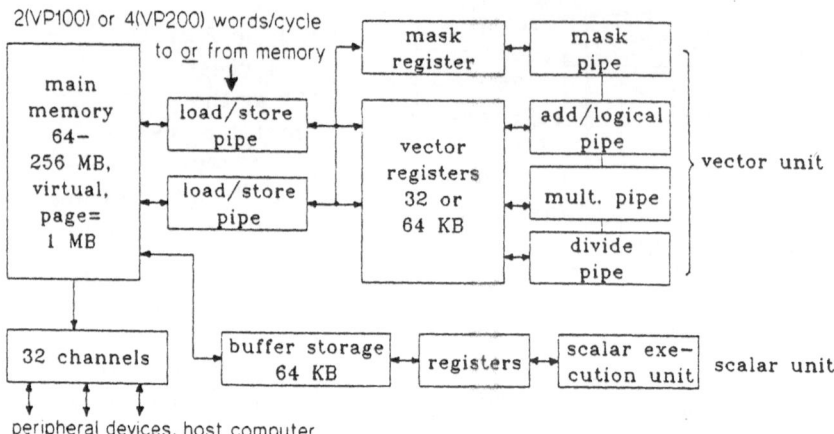

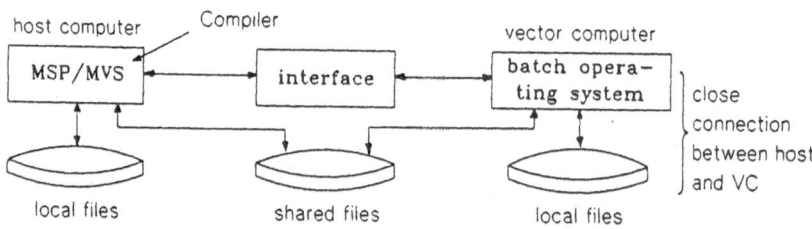

Fig. 2.4   Schematic view of the Fujitsu VP 100/200.

The VP 100 or 200 has also a bottleneck between main memory and vector registers. For the VP 100 resp. 200 only 2 resp.
4 words/cycle can be transfered, but for simple operations like (2.1) there are needed 3 resp. 6 transfers/cycle, i.e. there is available only 2/3 of the necessary bandwidth. It is interesting to note that the vector register file can be combined
dynamically to a few long or many short registers. There is a
close connection between host and vector computer: The compiler
runs only on the host computer and on the VP there are executed only the load modules.

Hitachi has developed on the same basic technology his
S810/10 and S810/20 and NEC will soon deliver its SX-2 with a
cycle time of 6 nsec, i.e. with an enhanced technology.

2.5 BOTTLENECKS

2.5.1 Between main memory and pipes

This bottleneck is closely related to the notion of supervector speed: To keep busy more than one pipe, i.e. to obtain more than one result per cycle time. The bottleneck may be caused by the lacking data paths to the main memory or by the lacking modularity of the main memory. It can be reduced for

2.5.1.1 the register-to-register machine by the chaining. If it is possible to overlap loading, execution and storing one can "uncouple" from the main memory and e.g. the add and mult pipes may be busy in parallel without (excessive) memory references.

2.5.1.2 the memory-to-memory machine CYBER 205 by the linked triad, e.g. for

$$c_i = a_i + s \times b_i , \qquad\qquad (2.2)$$

i.e. a triadic operation with one scalar operand s. There are needed only the available two load and one store operation from and to the memory. Then the add and mult pipes contained in each pipe of the CYBER 205 can be coupled together and work in parallel. Thus the linked triad is the "only chance" for supervector speed for the CYBER 205. If there would be a third load-path, then for the full vector triad also supervector speed could be obtained.

2.5.1 i/o-bottleneck between main memory and secondary storage units

For large 3-D problems not all the data fits into the main memory. Then in an out-of-core version of the program operands must be fetched from the secondary storage. In order to demonstrate the i/o-bottleneck we assume that in the simple operation (2.1) one of the two operands e.g. a_i must be fetched from the secondary storage and that b_i and c_i reside in the main memory. Then we can transform the measured MFLOPS-rate into a necessary transfer rate, neglecting the access times. The resulting transfer

Table 2.1 Measured MFLOPS-rates and corresponding transfer rates for one operand.

Computer	CRAY-1	CRAY X-MP one processor	CYBER 205 1-pipe	2-pipe	4-pipe
measured peak MFLOPS for +,× for n=10 000	23.4	67.5	49.7	98	197
corresponding bits/sec for one operand	$1.5 \cdot 10^9$	$4.3 \cdot 10^9$	$3.2 \cdot 10^9$	$6.3 \cdot 10^9$	$12.6 \cdot 10^9$

rates can be seen in Table 2.1. Now we compare these transfer rates to those of a single disk and for parallel i/o by 10 disk channels:

disk IBM-type 3350:$3.5 \cdot 10^7$ bits/sec, 10 disks:$3.5 \cdot 10^8$ bits/sec,
3380:$8.1 \cdot 10^7$ bits/sec, 10 disks:$8.1 \cdot 10^8$ bits/sec.

If we compare these transfer rates, even for 10 parallel disk
channels, we can see that they are not sufficient, and we have
still neglected the access times. Thus disks are an insufficient secondary storage medium for a vector computer for working
data sets.

If we now take as secondary storage medium the SSD (Solid
-state Storage Device) of CRAY and take the maximum number of
channels we can get the following transfer rates:

X-MP/1   100 MB/sec = $8.0 \cdot 10^8$ bits/sec,
CRAY-1   4×100 MB/sec = $3.2 \cdot 10^9$ bist/sec,
X-MP/2   1000 MB/sec = $8.0 \cdot 10^9$ bits/sec,
X-MP/4   2000 MB/sec = $16.0 \cdot 10^9$ bits/sec.

Comparison with Table 2.1 shows that we cannot meet the required transfer rate for the X-MP/1 because of only one channel
of 100 MB/sec. But for the other computers, the CRAY-1 and the
X-MP with 2 or 4 processors we can meet the requirements of
Table 2.1. We shall come back to the question of the i/o-bott-
leneck in section 4.3.

2.6 VISIBLE TRENDS IN DEVELOPMENT

The third Japanese vector computer, the NEC SX2 will be
officially announced during the printing of these proceedings.
It will have 6 nsec cycle time and will reach by 4-stream vector pipes (results in large $n_{1/2}$!) a peak rate of 1.3 GFLOPS
(gigaflops = 10^9 floating point operations per second).
A first report about the basic concept can be found in [3].

CRAY will announce officially also during the printing of
these proceedings its CRAY-2, with 4 nsec cycle time and 4 processors which are equipped like the CRAY X-MP/4 with hardware
instructions for gather/scatter, compressed index instructions
etc. The peak performance will be 2 GFLOPS. The main memory
will have 64 to 256 Mwords (!!) which will allow to treat really large problems incore. Later versions should have up to 16
processors. The operating system will be UNIX and not COS.

CDC has founded a new Company ETA for the development of
the successor of the CYBER 205. Until 1987 they will develop
the ETA^{10} with 5 nsec cycle time, up to 8 processors, each one
with two pipes. Each processor will have a local memory and
there will be a very large shared memory, the size depending
on the available technology. The maximal hardware vector length
of 65535 will be retained because of the compatibility to the
CYBER 205. The peak performance for 64 bit arithmetic will be
6.4 GFLOPS. The operating system presently is planned to be
VSOS and UNIX.

For the further development the CRAY-3 is expected with 1 nsec cycle time and the later computer of ETA should have 5/3 nsec cycle time. This results in a corresponding increase of the peak vector but also of the peak scalar performance. We think that around 1992 we shall approach the 100 GFLOPS which means the 1000 fold speed of the first generation of vector computers.

3. SOFTWARE

3.1 OPERATING SYSTEMS

The operating systems are rather simple and even if there is a time sharing possibility most of the computing centres do not allow this possibility to the users. Thus the simplicity of the operating system can be accepted within some reasonable limits. What we get for this simplicity is a relatively small overhead for the operating system. What we have to pay is that the resource management is not so sophisticated as in a general purpose computer, above all if there are several large jobs executing simultaneously. This might result in an increased necessity of operator activities.

3.2 COMPILERS AND FORTRAN 77

These two points are closely related to each other. The compilers are presently the weakest point of all vector computers and this is to a certain amount due to the fact that FORTRAN 77 has no vector possibilities. If we want to develop portable software, i.e. software which should run also on a general purpose computer or on other vector computers, we write the program in FORTRAN 77 and then use the <u>autovectoriser</u> which should transform the source code into vector operations whenever possible. But ultimately the autovectorisers analyse only the innermost do-loop, and they give up if in this loop there is only one nonvectorisable statement, instead of restructuring the loop.

The efficiency of the compiler can be measured by the comparison of the compiler-generated code to an assembler code for the same problem. Let us consider the matrix multiplication of two 100×100 matrices. In Table 3.1 there are presented the measured MFLOPS-rates for the optimal algorithm in FORTRAN 77 by the simultaneous calculation by columns, leading to contiguous vectors, and by an assembler coded program. The innermost do-loop has a linked triad as statement, but for the discussion of the results we must consider the <u>two</u> innermost do-loops which are of the type

```
      do 20 k = 1,n
      do 20 i = 1,n
   20   c(i,j) = c(i,j) + a(i,k) × b(k,j)
```
(3.1)

scalar for i-loop .

Table 3.1 Measured MFLOPS-rates for the multiplication
          of two 100×100 matrices.

Computer	CRAY-1 M	CRAY X-MP one processor	CYBER 205 1-pipe	2-pipe
FORTRAN 77	34	68	43	53
Assembler	149	190	53	74

If we now consider the CRAY's which are register machines,
we can see a tremendous increase in the MFLOPS-rates from FOR-
TRAN 77 to the assembler program. In FORTRAN 77 the CRAY Compiler recognizes the linked triad in the innermost do-loop (the
b(k,j) is a scalar for the i-loop), but for the CRAY-1 the
MFLOPS-rate is limited by the bottleneck between main memory
and V-registers and on the measured CRAY X-MP/22 the MFLOPS-
rate is determined by the bank conflicts of the two processors
accessing the 16 banks of the 2 Mwords main memory. In the assembler program information of the outer k-loop is used: If we
do not process the whole range of the index i in (3.1) but only 64 elements (the length of the V-registers) then only a(i,k)
must be exchanged because the c(i,j) do not depend on k. Thus
instead of processing a whole column, each column is processed
in sections of 64 elements and then c(i,j) remains fixed in a
V-register. Instead of two loads for c(i,j) and a(i,k) and one
store for c(i,k) then there is only one load for a(i,k). This
demonstrates clearly what we meant with the great chance and
danger of the register-to-register machine. If one can succeed
by the chaining to uncouple the operations as far as possible
from the main memory, supervector speed can be obtained. But
it is at the same time a severe problem for the compiler to do
an optimal chaining.

Thus we consider it as the essential deficiency of the
CRAY compiler that it analyses only the innermost do-loop. This
means that there is not possible a "global" chaining as we have
seen it above, nor there is possible a loop restructuring. Also
some constructs can be recognized as non-recursive, only in
a global view. CRAY tries to vectorise if's as far as possible.
As long as there is still a considerable gain by assembler programming, above all by "hand-chaining", the compiler has not
yet reached a "satisfactory" state.

If we consider in Table 3.1 the memory-to-memory machine
CYBER 205 we can see that there is not so much gained by the
assembler program. The reason is that for the "simpler" hardware the FORTRAN 77 compiler can already generate a relatively
optimal code, in this case for the linked triad. An assembler
program can gain only by reducing the overhead for the storage
access to get the b(k,j) in (3.1). If the innermost do-loop is
unrolled 5-fold and at the beginning of this unrolled part 5
values b(k,j) are fetched by one operation and stored into 5

locations of the register file the overhead is reduced corre-
spondingly and we get the increased MFLOPS-rate of Table 3.1.
Here also information of the outer k-loop has been used.

The philosophy of CDC for the CYBER 205 was not primarily
directed towards portable code. CDC has rather invented language extensions to FORTRAN IV which allow an explicit vectorisation and the vectorisation of conditions by a where-statement.
These language extensions result in an excellent use of the
hardware, but the portability is lost! This philosophy is the
reason that the autovectoriser has only limited possibilities:
there are no compiler directives, e.g. to suppress vectorisation of short do-loops, there is no vectorisation of if's. The
user has to take himself the responsibility by the "unsafe"
option that the vector length is less than 65536. Many loops
are not vectorised because "suspicion of recursion".

There are preprocessors VAST (Pacific Sierra Research
Corporation) and KAP (D.Kuck) which translate standard FORTRAN
to the special CYBER 205 - FORTRAN. But a comparison of these
preprocessors and the CDC autovectoriser by C.Arnold [4] reveals that none of the autovectorisers is ideally suited. D.
Kuck [5] has developed a theory by his data dependency graph
to autovectorise language constructs as far as possible, but
the reality obviously still presents problems. So we consider
the problem of autovectorisation as not yet solved. It is interesting to note the policy of the ETA corporation: Their basic compiler will have no autovectoriser, but they will deliver preprocessors which should accomplish the autovectorisation.

In some seminars or conferences, e.g. in [6], the Japanese pretended that their compilers autovectorise better than
previous compilers. But as the relevant computers are register-
to-register machines one has to wait for practical experiences,
above all concerning the chaining. If there is no global chaining then there will remain the discrepancy between compiler
generated and assembler coded programs.

At the end of this section about compilers and autovectorisation there should be expressed the following

Warning: No autovectoriser or preprocessor can produce from a
"bad" program a "good" vector program. You must devise good
programs yourself!

3.3 SOME REMARKS TO FORTRAN 8X

Presently extensions to FORTRAN are under discussion as
FORTRAN 8X [7,8,9,10]. These extensions would offer
to the user the possibility of writing (more) portable code,
to the compiler to vectorise much more efficiently,
to the manufacturer to invent hardware instructions which efficiently realize the new language constructs.
In the following only briefly the main features of these extensions will be reviewed. But this will be sufficient to demon-

strate where the essential progresses are to be expected.

One feature is the extension of operators and intrinsic functions to arrays, thus also to vectors. One may write down simply

$$e = a * b + c * sin(d), \qquad (3.2)$$

which looks like FORTRAN 77, but now all the operands are arrays of the same dimension and the operations are executed <u>by elements</u>. This feature saves a lot of "unnecessary" do-loops and allows an efficient vectorisation above all for multidimensional arrays. There can be selected "array sections", e.g.

$$a(:,j) \qquad (3.3)$$

selects the j-th column of the matrix a.

A new statement is the where-statement which can be used in the form

```
where (a.gt.0)    or    m = a.gt.0
   x = u*v/w              where (m)
otherwise                   .                        (3.4)
   x = z*w/u                .
end where   .               .
```

In the second form a logical (mask) vector m has been created first and then is used in the where-statement. All operations are again array operations in the sense of (3.2). <u>All</u> elements for the assignments in the where and otherwise branch are computed and then selected according to the logical expression in the parenthesis of the where statement. Only assignment, pack and unpack are allowed in the where statement. Thus one can write

$$where(m)\ pack(au,u,n)\quad or\quad where(m)\ unpack(u,au,n)\ .\quad (3.5)$$

The pack statement selects the elements of a vector u into a dense vector au where the mask vector m is true (naturally m and u must have equal dimension) and the variable n (integer) then gives the number of selected elements. The inverse operation is the unpack, where the elements of au are distributed into the elements of u according to the mask vector m. These operations could be called gather/scatter by mask vector m. By the where statement many if's can be avoided and replaced by "vectorisable" conditions.

In many applications, e.g. in finite element calculations, indirect addressing cannot be avoided, e.g. we have b(k) = a(i(k)) and a(i(k)) = b(k). In FORTRAN 8X this is expressed simply by

$$b = a(i)\quad and\quad a(i) = b\ .\qquad (3.6)$$

Here i is an integer <u>vector</u> of the indices to be selected and has the same dimension as b. One has to think a moment about

14

the power of a statement like (3.6). This could be called ga-
ther/scatter by index list. There will be many new intrinsic
functions, e.g.

c = matmul (a,b)   matrix multiplication C=A·B,

                                                            (3.7)

s = dotproduct(a,b)   scalar product of vectors a,b .

    We now recognize that it will be much easier for the user
and for the compiler to generate much better and portable vector programs. J.Reid, who feels responsible in the X3J3 Committee for the language extensions to support efficient vectorisation, would appreciate comments or suggestions to such extensions (J.K.Reid, AERE Harwell, Oxfordshire OX11 ORA, Great
Britain).

4. SOME EXPERIENCES OF THE NUMERICS
GROUP AT THE COMPUTER CENTRE OF
THE UNIVERSITY OF KARLSRUHE

4.1 BACKGROUND

    In the Numerics Group at the Computer Centre (Rechenzen-
trum) of the University of Karlsruhe we develop a program
package FIDISOL (finite difference solver) for a selfadaptive
variable order/variable step size finite difference method for
the solution of nonlinear systems of elliptic and parabolic
2-D and 3-D PDE's on a rectangular domain. This "black box"
software should be portable as far as possible, it is tested
on general purpose computers and on the vector computers CRAY-
1M, CRAY X-MP/2, CYBER 205 and (soon) on the VP 200. The
"dream" which we want to realize is sketched below:

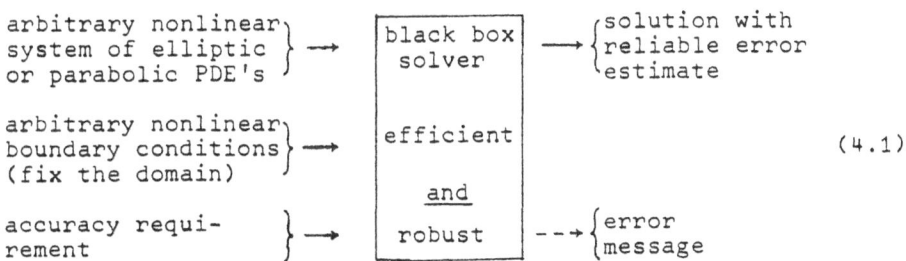

This "black box solver" should be efficient and robust, but
<u>highly efficient</u> means fully adapted to a special problem,
<u>robust</u> means, that the solver works for "all types" of
problems.
There is quite naturally a contradiction between these two requirements. Therefore general purpose software is always a <u>com-</u>
<u>promise</u> between efficiency and robustness. This holds still

more pronounced if vectorisation is included.

In the following subsections we report about experiences which we have gained in the course of this software development and which we think to be of rather general interest. It would be impossible to describe in this paper the whole development.

4.2 DATA FLOW ALGORITHM

The key to an efficient vectorisation is an optimal data structure which allows a continuous processing of the data. Theorists develop the ideas of data flow supercomputers which are in contrast to the usual computers not program-controlled but data-controlled, see e.g. Dennis [11]. But we must solve our problems of CFD on existing vector computers. What we can learn from these theorists is the priority of a continuous data flow. This led us to the pragmatic definition [12] of a

data flow algorithm: Prescription for the solution of a problem where the structure of the data and the sequence of operations is controlled by the absolute priority of a continuous processing of the data as far as possible on the real architecture of a given computer.

The key to a data flow algorithm is the separation of the selection and of the processing of the data. A construct of the type
        if ... (selection)   then ... (processing)
means mixing of data selection and data processing and thus prevents an efficient vectorisation. Therefore if we need for different tasks different data structures we must transform the data to the required structure by sorting or merging before we execute the task. We want to demonstrate this for the following

example: Computation of 3-D derivatives and of corresponding error estimates.

We assume that we want to compute the second and first derivatives u_{xx} and u_x and its error estimates d_{xx} and d_x. If we solve a system of n PDE's the solution vector u has n components, ℓ denotes one component:

$$u = \begin{pmatrix} u_1 \\ \vdots \\ u_\ell \\ \vdots \\ u_n \end{pmatrix} \longleftarrow \text{component } \ell \ . \qquad (4.2)$$

For each component ℓ e.g. the second derivative is computed by the evaluation of a difference formula of the type

$$u_{\ell,xx} = \sum_m a_m \, u_{\ell,m} \ . \qquad (4.3)$$

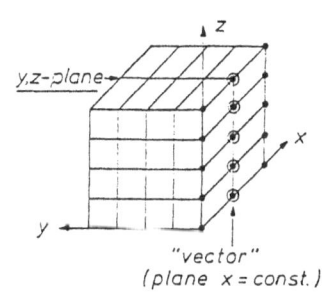

y,z-plane

z

x

y

"vector"
(plane x = const.)

Fig. 4.1 Illustration to
the evaluation of the
difference formula (4.3).

The same formula holds for all
points of a plane x = const, see
Fig. 4.1. In (4.3) m denotes a
point of one of the planes x_m = const.
This situation suggests the
<u>simultaneous</u> computation for u_{xx}
for all points of a plane
x = const. This is executed by the
following nested loop:

for all components ℓ do

  for all planes x_i do   (4.4)

    for all planes m of formula do

      for all points p of plane m do

      uxx(p,i,ℓ) = uxx(p,i,ℓ) +

        + a(i,m) * u(p,g(i,m),ℓ)

where g(i,m) selects the actual plane. If the solution component u_ℓ is sorted in <u>planes x = const</u> (4.4) is a data flow algorithm which runs over the contiguous "vector" of all u_ℓ of a
plane x = const, see Fig. 4.1.

If we want to compute in a similar way the y- and z-deri-
vatives and error estimates we need a corresponding sorting of
the data in planes y = const and z = const. Such a sorting is
always a periodic sorting (constant stride). We want to demonstrate this in the following.

If ℓ denotes one of the n components of the solution of a
system of n PDE's and i,j,k are the grid indices for x_i, y_j,
z_k, the solution will be delivered from the difference
method in the sorting $u_{\ell,i,j,k}$. The first step then will be to
separate the components, i.e. the index ℓ should be the last one
$u_{\ell,i,j,k} \to u_{i,j,k,\ell}$ which means to take every n-th element. This
is illustrated in the following scheme, taking n = 3 and deno-
ting the three components by f,g,h:

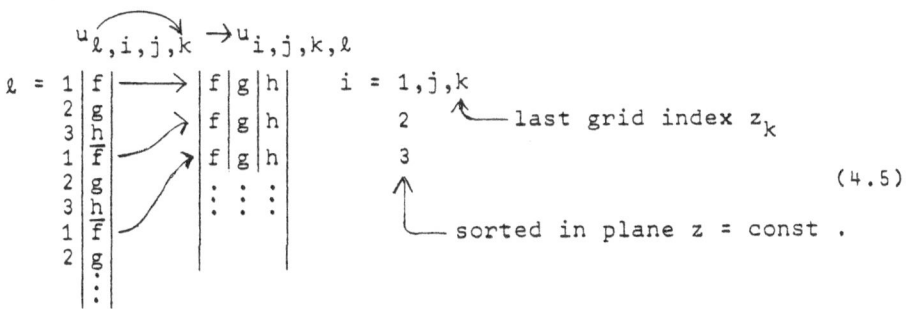

The result $u_{i,j,k,\ell}$ is sorted in planes z_k = const because k is
the last grid index. For sorting in planes

17

x_i = const i must be the last grid index. This is accomplished in the following way:

Sorting for plane x_i = const. $u_{i,j,k,\ell} \rightarrow u_{j,k,i,\ell}$
(separately for each component ℓ)

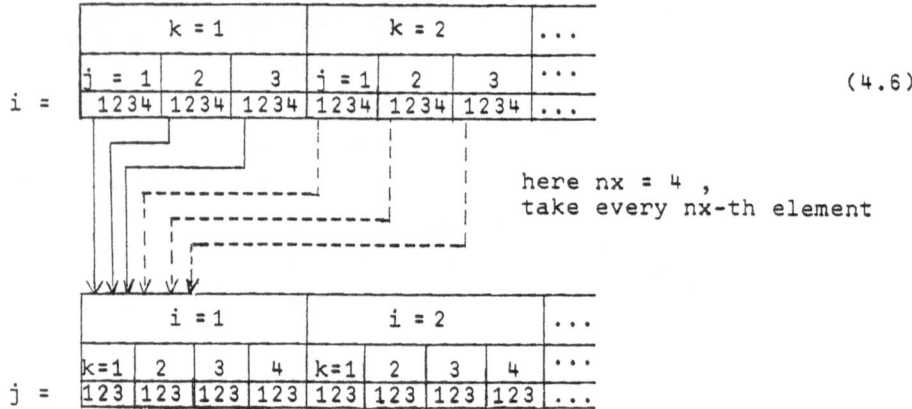

(4.6)

here nx = 4 ,
take every nx-th element

If the number of grid lines is nx the simple rule is again: Take every nx-th element. By the sortings

$$\rightarrow u_{i,j,k,\ell} \rightarrow u_{j,k,i,\ell} \rightarrow u_{k,i,j,\ell}$$ (4.7)

we have the solution in planes z,x,y = const. By taking every nx-th, ny-th, nz-th element we transform one sorting into the other.

Does this sorting pay? It pays always. Processing with constant stride means gather/scatter operations for the CYBER 205 and the danger of bank conflicts for the CRAY's. For separate selection only the <u>selection</u> is retarded, processing then is with full speed. This is our main message concerning data flow algorithms.

4.3 THE I/O-BOTTLENECK

To discuss the problem of the i/o-bottleneck we consider our
<u>typical example</u>: In fluid dynamics we have a system of 6 3-D PDE's, a 50×50×50 grid and a 4th-order difference star. Then we have

     750 000 unknowns and circa
     59 million nonzero elements in the   (4.8)
     matrix of the resulting linear system .

The matrix has the structure of scattered diagonals. Therefore we use <u>diagonal storing</u> for the matrix: it is stored as a "file" of diagonals. But even if this seems to be an ideal storing there results a considerable storage overhead. Because we have a system of PDE's there are block diagonals and storing them as real diagonals means filling in of zeros in the triangles between the blocks, see Fig. 2 in [12]. Similarly the higher order formulae create near the boundaries additional sparse diagonals which must be filled up by zeros. So we end up with 291 diagonals which need circa 220 million storage locations for the circa 59 million nonzero elements.

For the solution of such large sparse linear systems only iterative methods can be used because direct methods must be excluded because of the fill-in. Compatible to diagonal storing are iterative methods which are based on the matrix-vector-multiplication (MVM), e.g. CG-type methods. For the computation of c = A·r the vectors r and c are kept in the main storage and the matrix A is shifted and processed diagonal by diagonal in the main storage, see Fig. 3 in [12]. If we assume that we have a 4-pipe CYBER 205 with a peak rate of 200 MFLOPS and if we assume that we need 600 MVM for the solution of the problem, then the necessary CPU-time is

$$\text{CPU-time} = \frac{\overset{\text{diag.}}{291} \cdot \overset{\text{elements}}{750\ 000} \cdot \overset{\text{MVM}}{600} \cdot \overset{\text{oper}}{2}}{\underset{\text{oper/sec}}{2 \cdot 10^8}} = 1310\text{sec} = \underline{21.8\text{min.}} \qquad (4.9)$$

For each element there are two operations, a multiplication and an addition.

The corresponding i/o-time is, if we only consider the transfer rate of <u>one</u> disk channel of IBM 3350-type and neglect the access times:

$$\text{i/o-time} = \frac{\overset{\text{diag.}}{291} \cdot \overset{\text{words}}{750\ 000} \cdot \overset{\text{readings}}{600} \cdot \overset{\text{bits}}{64}}{\underset{\text{bits/sec}}{3.5 \cdot 10^7}} = 239451\text{sec} = 66.5\text{h(!)}$$
$$(4.10)$$

If we consider an algorithm which allows to execute 2 MVM with one reading of the matrix (the biconjugate gradient algorithm, see [13]), we can reduce the i/o-time to

$$\text{i/o-time} = 0.5 \cdot 66.5\text{h} = \underline{33.25\text{h}} \ . \qquad (4.11)$$

If we distribute each diagonal (750 000 elements) onto 10 separate disks which we call <u>file banking</u> because the different parts of a diagonal are stored on different files like on different banks in a main memory, we can execute parallel i/o by 10 disk channels and thus reduce the i/o-time to

$$\text{i/o-time} = 0.1 \cdot 33.25\text{h} = \underline{3.33\text{h}} \ . \qquad (4.12)$$

If we now throw out the old disks and put into the computing centre the new IBM-3380-type disks with $8.1 \cdot 10^7$ bits/sec transfer rate we can reduce the i/o-time to

$$\text{i/o-time} = \frac{3.5}{8.1} \cdot 3.33h = \underline{1.44h} \ . \tag{4.13}$$

The next possibility is packed storing of the diagonals: The sparse diagonals are stored by a bit vector and the nonzero elements. This measure reduces the 220 million storage locations to circa 70 million storage locations. Then the i/o-time is reduced to

$$\text{i/o-time} = \frac{70}{220} \cdot 1.44h = 0.458h = \underline{27.5min} \ . \tag{4.14}$$

But the diagonals must be unpacked before processing which means one additional operation for every four operations. Thus 25% of the original CPU-time is wasted for reducing the i/o and the CPU-time is increased to

$$\text{CPU-time} = (5/4) \cdot 21.8min = \underline{27.3min} \ . \tag{4.15}$$

Thus by the combination of all these measures (we use 10 new disk channels!) we could just balance the CPU- and i/o-time. But most of the installed vector computers do not have so many (new) disks. Thus the i/o-bottleneck limits in many problems a full use of the vector computer. Disks are in no case a sufficiently fast secondary storage medium for a vector computer. This practical example should once more stress this problem which has been illustrated already in section 2.5.1.

If we now consider CRAY's SSD for the same problem, the i/o-time is (we neglect again the access times)

$$\text{i/o-time} = \frac{\overset{\text{words}}{70 \cdot 10^6} \cdot \overset{\text{readings}}{300} \cdot \overset{\text{bits}}{64}}{\underset{\text{bits/sec}}{8 \cdot 10^9}} = 168sec = 2.8min(!) \tag{4.16}$$

This illustrates the usefulness of a fast secondary storage medium.

But the best solution would be a sufficiently large main memory. For the present generation of vector computers we consider a main memory of 128 Mwords to be adequate. For the next generation the size should be 128 times the speed-up factor against the present generation. So we consider the 256 Mwords of main memory for the CRAY-2 not sufficient, because with the increased speed the problem size and thus the amount of data grows considerably. This speed can be used reasonably only for incore data. Thus also for a CRAY-2 with 256 Mwords main memory a type of SSD will be necessary.

4.4 THE PROBLEM OF PORTABLE SOFTWARE FOR DIFFERENT VECTOR COMPUTERS

Full efficiency of a program means adaptation of the program to the special vector computer, but this contradicts the requirement of portability. What to do then? Today one must adapt a program to each type of vector computer, else too much efficiency is lost. This situation will hold on until satis-factory compilers with satisfactory autovectorisers will be available for FORTRAN 8X. But what about portability?

Our compromise for this problem is to maintain a standard version for FORTRAN 77 and to concentrate all machine-specific properties in "modules" which are subroutines or functions. Thus adaptation to a specific computer means exchange of modu-les and not of statements and thus allows a much better main-tenance. Presently the FORTRAN 77 library contains 107 modules. In the CYBER 205 library there are 34 modules which differ from the standard modules and in the CRAY library there are (only) 15 differing modules. For the generation of a specific version the necessary source code modules are selected and com-bined in a new library by corresponding job control statements. In the specific CYBER 205 modules there are Q8-calls or expli-cit vector operations above all for loops which are not vecto-rised by the autovectoriser because of suspicion of recursion. In the specific CRAY modules there are mostly calls of optimi-sed assembler routines, e.g. SDOT or ISMAX. For the CRAY there is a further special version with 3 additional modules which has unrolling of do-loops in the matrix-vector-multiplication.

One serious problem for the portability is the i/o. Also these i/o activities are contained in machine-specific modules which allow a flexible adaptation to the existing and future i/o possibilities. For the CYBER 205 implicit i/o is used and for the CRAY's we use "buffer in" and "buffer out" for sequen-tial access and word-addressable i/o for direct access.

To the user we offer presently 3 possibilities for the MVM which he can choose by parameter setting in the control vector, but which need eventually additional JCL statements, e.g. for several i/o channels:
1. Normal i/o, diagonals in full length, one disk channel.
2. Packed diagonals, one disk channel.
3. Packed diagonals plus file banking (several disk channels).
But the number of versions then is = number of methods × number of vector computers, which gets quickly a rather big value.

4.5 EXPERIENCES WITH ILU-PRECONDITIONING ON A
    VECTOR COMPUTER

The biconjugate gradient (BICO) method is an efficient generalisation of the CG (conjugate gradient) method. Practi-cal experiences on a vector computer have been reported in [13]. It is "well known" that preconditioning can essentially improve the efficiency of CG-type methods. ILU is the incomple-te LU-decomposition, see e.g. H.van der Vorst [14], which exists in different versions. The basic idea is to execute a LU-decomposition, but to "forget" all elements which drop out-side of the already existing diagonals of the original matrix. Thus fill-in is avoided, but the result is only an approxima-tion to the true solution. Nevertheless this ILU can be used as a preconditioner to a CG-type method.

21

The idea of preconditioning is the following: In BICO e. g. the essential operation is a matrix-vector-multiplication $A \cdot r$ (and $A^T r$) where r is a "search direction" and A is the matrix of the linear system. If one uses a "better" search direction $B \cdot r$, where B is a "preconditioning matrix", one may get a faster convergence. It turns out that if $B = A^{-1}$ one gets immediately the solution. Thus the optimal search direction is

$$c = Br = A^{-1}r \implies Ac = r . \qquad (4.17)$$

Thus one has to solve the same type of equation to get c as the original system. But one can solve $Ac=r$ only "approximately", e.g. by ILU. Table 4.1 shows the results which

Table 4.1 Results for BICO and ILU-preconditioned BICO on a 2-pipe CYBER 205.

method	number of iterations	CPU sec scalar	CPU sec vector	gain $g = \dfrac{\text{CPUscalar}}{\text{CPUvector}}$
BICO	279 ⌉ 3.4	21.2 ⌉ 1.6	1.50 ⌉ $\frac{1}{3.2}$	14.1
ILU-BICO	81 ⌋	13.4 ⌋	4.8 ⌋	2.8

we obtained for a 2-D test-PDE and 67×67 grid on a 2-pipe CYBER 205 with diagonal storing of the matrix. ILU reduces the number of iterations by a factor of 3.4. Because each iteration step with ILU preconditioning is more expensive the CPU-time in scalar mode is reduced only by a factor 1.6, but nevertheless it is reduced which corresponds to the "well known" experiences on general purpose computers. But in vector mode the CPU-time goes up by a factor 3.2. The reason is visible by the vectorisation gain g: BICO is 14.1 times faster in vector mode than in scalar mode, i.e. we have an excellent vectorisation with long vectors, ILU-BICO has only a poor vectorisation with g=2.8, i.e. the recursive character of ILU-preconditioning has destroyed the high vectorisation gain and thus inverted the scalar experiences. Similar experiences are reported in [15].

The conclusion which we can draw from this numerical experiment is that experiences which have been gained on general purpose computers can be inverted on a vector computer if the vectorisation gain is considerably reduced by poor vectorisation.

These few points which we reported from our experiences in developing "black box" software, which should run efficiently on different vector computers, demonstrate the additional difficulties which we encounter if the problem of vectorisation comes into the play. We think that the users, but also the compiler designers and the hardware designers are still in an "early" phase of learning to use and to design vector computers.

5. NUMERICAL EXAMPLES OF FIDISOL

In this section we report about the solution of test problems with the FIDISOL (finite difference solver) program package which is at the time of writing this paper in the final state of development. The essential idea is that together with the solution there is computed an error estimate. Further information about the solution method is given in [16,17]. In this paper we present some results to demonstrate the possibilities of such software on different vector computers.

A first test example is an artificial 3-D system of 3 PDE's which is a rough simulation of the Navier-Stokes equations:

$$u_{xx}+u_{yy}+u_{zz}+R\cdot(uu_x+vu_y+wu_z)+u+h_1 = 0 \ ,$$
$$v_{xx}+v_{yy}+v_{zz}+R\cdot(uv_x+vv_y+wv_z)+v+h_2 = 0 \ , \qquad (5.1)$$
$$w_{xx}+w_{yy}+w_{zz}+R\cdot(uw_x+vw_y+ww_z)+w+h_3 = 0 \ .$$

The functions h_1 - h_3 are adapted to prescribed solutions u,v, w which are trigonometric functions in x,y,z. We prescribe Dirichlet boundary conditions (i.e. the values of u,v,w) on the surfaces of the unit cube. We use a 4-th order method on an 18×18×18 grid. The parameter R which has the meaning of a "Reynoldsnumber" has the value R=50. In Table 5.1 are presented the

Table 5.1   CPU-time for the solution of equ.(5.1) on different vector computers.

computer	mode	CPU-time	gain g
CRAY X-MP/22 (one processor)	scalar	84.2 sec	8.59
	vector	9.8 sec	
CRAY-1M (1 Mword)	scalar	115.3 sec	5.36
	vector	21.5 sec	
CYBER 205 (2-pipe,1 Mw)	scalar	152.4 sec	15.71
	vector	9.7 sec	

timings for different vector computers. For the CRAY's the "standard CRAY-version" (without unrolling of loops for the MVM) has been used. Because the vector length runs over all unknowns (18×18×18×3 = 17496) the CYBER 205 gives an excellent timing in vector mode and an excellent gain g = 15.71.

As we know the solution of (5.1) we know also the exact error as the difference of the exact and the numerical solution. The maximal relative (to the maximum of the solution) error is o.19·10^{-4} and the error estimate is also 0.19·10^{-4}. We consider this property of the FIDISOL program package, to deliver together with the solution an estimate of the error, as the most valuable property. Then the engineer knows something about the numerical error of his solution. Thus he is able to separate the "model errors", which have been introduced by simplifying his problem to the system of (mostly nonlinear) PDE's and BC's, from the numerical errors. The error estimate is a local estimate for each grid point. Thus the user of FIDISOL can "see" where the errors are large. In the final version there is included a selfadaptive process of the selection of grid and order, see [18].

The next example is the 2-D heat driven cavity. The equations are formulated in dimensionless quantities in velocity vorticity formulation: u,v are velocity components, $\omega = v_x - u_y$ is the vorticity, T is a temperature disturbance:

$$u_{xx} + u_{yy} + \omega_y = 0 ,$$
$$v_{xx} + v_{yy} - \omega_x = 0 ,$$
$$u\omega_x + v\omega_y - Pr(\omega_{xx} + \omega_{yy}) - RaPrT_x = 0 ,$$
$$uT_x + vT_y - (T_{xx} + T_{yy}) = 0 .$$

(5.2)

Ra and Pr are Rayleigh and Prandtl number. The first two equations result from the continuity equation and definition of the vorticity by differentiating w.r.t. x and y and eliminating the mixed derivatives. The third equation results from the Navier-Stokes equations with Boussinesq approximation by eliminating the pressure, and the fourth equation is the energy equation. A rather sophisticated nondimensionalisation has been used, see [19]. The boundary conditions on the surface of the unit square are:

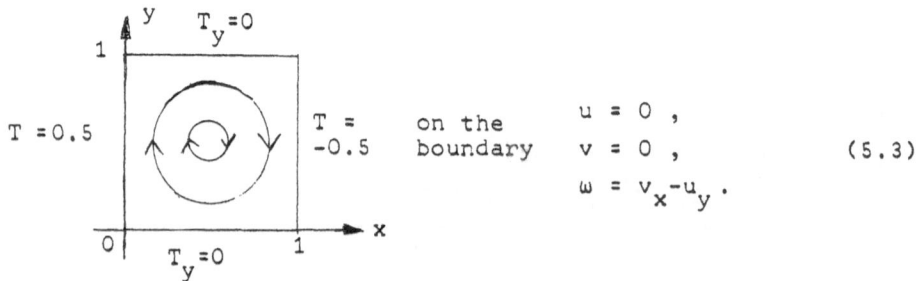

$$
\begin{aligned}
u &= 0 , \\
v &= 0 , \\
\omega &= v_x - u_y .
\end{aligned}
$$

(5.3)

The temperature boundary conditions are for the left/right boundary heated/cooled and the lower and upper boundary isolated. There results a convective flow as indicated in (5.3).

We have used the flow parameters Ra = 10^3, Pr = 0.71, a 40×30 nonequidistant grid and the orders 6 in x,y-direction. We measured the CPU-time for the following computers:

CRAY X-MP/22 (one processor)	14.5 sec ,		
CRAY-1 M (1 Mword)	33.9 sec ,		(5.4)
CYBER 205 (2-pipe, 1 Mword)	17.8 sec .		

For the CRAY's an optimised version with unrolling of loops in the MVM has been used. The estimated max. relative errors and the number of matrix-vector-multiplications (MVM) is presented in Table 5.2. It is

Table 5.2 Estimated max. relative errors for the
          solution of (5.2), (5.3) and number of
          MVM on different vector computers.

Computer	CRAY X-MP	CRAY-1	CYBER 205
error of u	$1.3 \cdot 10^{-4}$	$1.2 \cdot 10^{-4}$	$1.7 \cdot 10^{-4}$
v	$1.5 \cdot 10^{-4}$	$1.2 \cdot 10^{-4}$	$1.4 \cdot 10^{-4}$
ω	$6.1 \cdot 10^{-4}$	$6.1 \cdot 10^{-4}$	$6.2 \cdot 10^{-4}$
T	$1.6 \cdot 10^{-4}$	$1.4 \cdot 10^{-4}$	$1.5 \cdot 10^{-4}$
number of MVM	1976	2064	2016

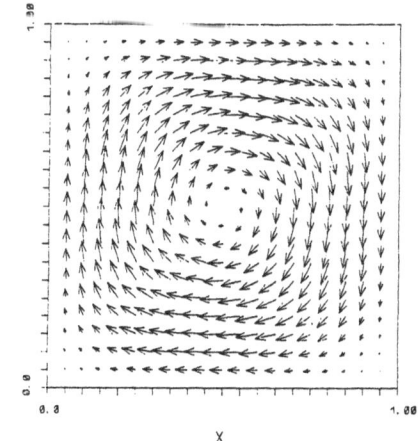

Y

0.0
0.0   1.00

X

Fig. 5.1 Velocity vector plot
         of the solution of
         (5.2), (5.3).

interesting to see that the error estimates and the MVM differ slightly on the different vector computers because of the different arithmetics (even on the CRAY's). In Fig. 5.1 there is depicted the velocity vector field. In Fig. 5.2 there are plotted contour lines of the stream function ψ which has been computed (also with FIDISOL) from

$$\psi_{xx} + \psi_{yy} + \omega = 0 ,$$
$$\psi = 0 \quad \text{on boundary.} \tag{5.5}$$

As we mentioned at the last example the computation of the errors allows to "see" where they are concentrated (in Table 5.2 there are

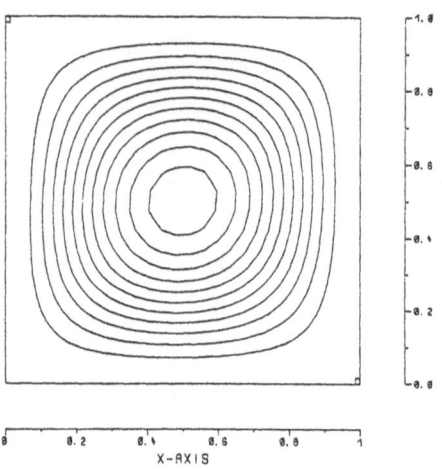

Fig. 5.2 Contour plot of the
stream function ψ.

presented only the max. errors). Therefore in Fig. 5.3 there are visualised (the scale does not matter) the contour lines for the variables u, v, ω, T and their relative errors (relative to the max. of the corresponding solution component). The errors of the vorticity ω are concentrated near the corners. It might be clear from this figure that only an automatic mesh and order selection process is able to find the optimal grid and order because each solution component has its maximal error at another position. Such a process is described in [18].

The last example is the 3-D convective flow in a heat driven cube. We use again a velocity-vorticity formulation with the velocity vector $\underline{v}$ and the vorticity vector curl $\underline{v}$:

$$\underline{v} = \begin{pmatrix} u \\ v \\ w \end{pmatrix} , \quad \text{curl } \underline{v} = \begin{pmatrix} \xi \\ \eta \\ \zeta \end{pmatrix} = \begin{pmatrix} w_y - v_z \\ u_z - w_x \\ v_x - u_y \end{pmatrix} . \tag{5.6}$$

T is a temperature disturbance. The system of 7 PDE's for the 7 unknown functions is

$$u_{xx} + u_{yy} + u_{zz} - \eta_z + \zeta_y = 0 ,$$
$$v_{xx} + v_{yy} + v_{zz} + \xi_z - \zeta_x = 0 ,$$
$$w_{xx} + w_{yy} + w_{zz} - \xi_y + \eta_x = 0 ,$$
$$-\xi u_x - \eta u_y - \zeta u_z + u\xi_x + v\xi_y + w\xi_z - Pr(\xi_{xx} + \xi_{yy} + \xi_{zz}) - RaPrT_y = 0 , \tag{5.7}$$
$$-\xi v_x - \eta v_y - \zeta v_z + u\eta_x + v\eta_y + w\eta_z - Pr(\eta_{xx} + \eta_{yy} + \eta_{zz}) + RaPrT_x = 0 ,$$
$$-\xi w_x - \eta w_y - \zeta w_z + u\zeta_x + v\zeta_y + w\zeta_z - Pr(\zeta_{xx} + \zeta_{yy} + \zeta_{zz}) = 0 ,$$
$$uT_x + vT_y + wT_z - (T_{xx} + T_{yy} + T_{zz}) = 0 .$$

The first three equations result from the continuity equation and the definition of the vorticity, the next three equations result from the Navier-Stokes-Equations with Boussinesq approximation by the elimination of the pressure, and the last equation is the energy equation. In Fig. 5.4 are sketched the boundary conditions. The two front sides of the depicted cube are hot and cold with a prescribed temperature distribution,

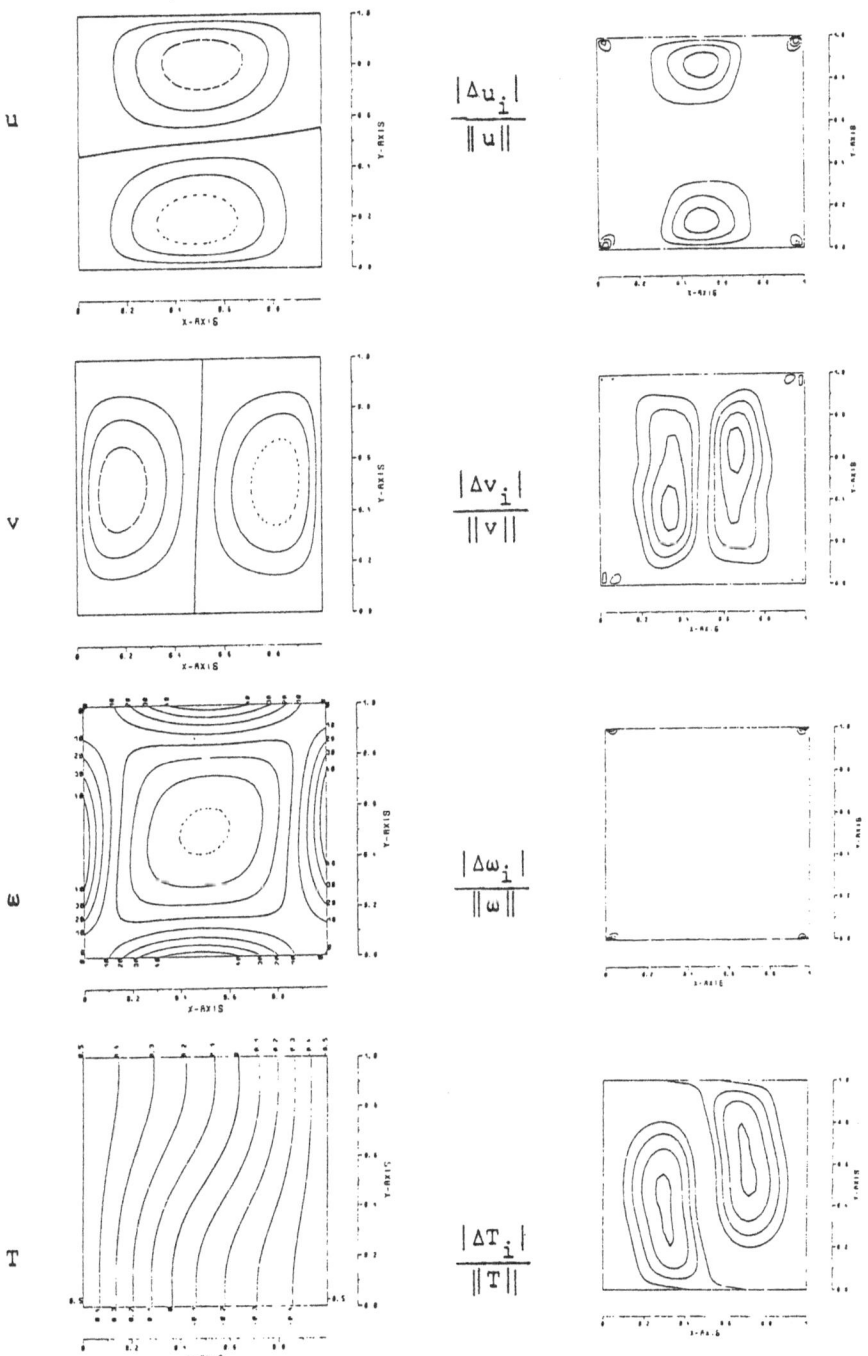

Fig. 5.3 Contour plots of the variables u,v,ω,T and of their
relative errors.

the remaining surfaces are isolated. The boundary conditions on the surface of the cube are:

$$u = v = w = 0, \quad \xi - w_y + v_z = 0, \quad \eta - u_z + w_x = 0,$$
$$\zeta - v_x + u_y = 0 .$$

On $y = 0$: $T = -0.5 \sin (\pi/2)x$, \hfill (5.8)

on $x = 0$: $T = 0.5 \sin (\pi/2)y$,

remaining surfaces: normal derivative

$$T_x = 0 \text{ or } T_y = 0 \text{ or } T_z = 0 .$$

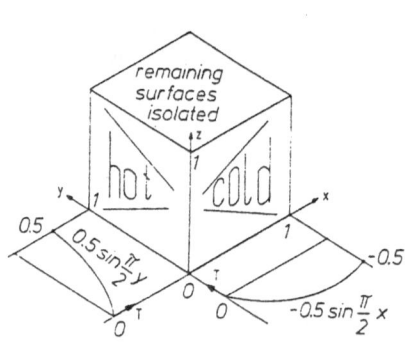

Fig. 5.4 Boundary conditions for T.

The system (5.7), (5.8) has been solved on the 2-pipe CYBER 205 (1 Mword) with the parameters $Ra = 10^3$, $Pr = 0.6991$ on a 15×15×15 equidistant grid with the orders 4 in each coordinate direction. There are 23625 unknowns. The CPU-time was 145.2 sec and there were needed 1246 MVM. The maximal relative errors of the variables were estimated to

$$u : 1.3\%, \quad \xi : 3.0\%, \quad T : 0.1\%$$
$$v : 1.3\%, \quad \eta : 3.0\%,$$
$$w : 1.2\%, \quad \zeta : 4.8\%, \hfill (5.9)$$

It is difficult to visualise 3-D results because it is not possible to draw a contour plot over a 3-D coordinate system. And in 3-D for 7 variables there is a vast amount of data. Therefore we present here only a few illustrations of the results. In Fig. 5.5 there are depicted velocity vector plots for different planes. Note the perspectivic view also of the velocity vectors. These plots illustrate comprehensively the flow. Note that near the hot and cold vertical walls there are larger velocities upwards and downwards than near the isolated walls. In Fig. 5.6 there is presented the contour plot of the vertical velocity w in the mid plane $z = 0.5$, illustrating how the flow recirculates up and down near the hot and cold wall (in this plane!). In Fig. 5.7 there are depicted the contour lines of the temperature disturbance (difference to the medium temperature in nondimensional scale). In the upper left and lower right corner there are the hot and cold peak values of T as prescribed. On the isolated surfaces the temperature distribution is induced by the convective flow. Note that this holds only in this mid plane.

In conclusion we can say that these examples have demonstrated that it is possible to develop efficient "black box"

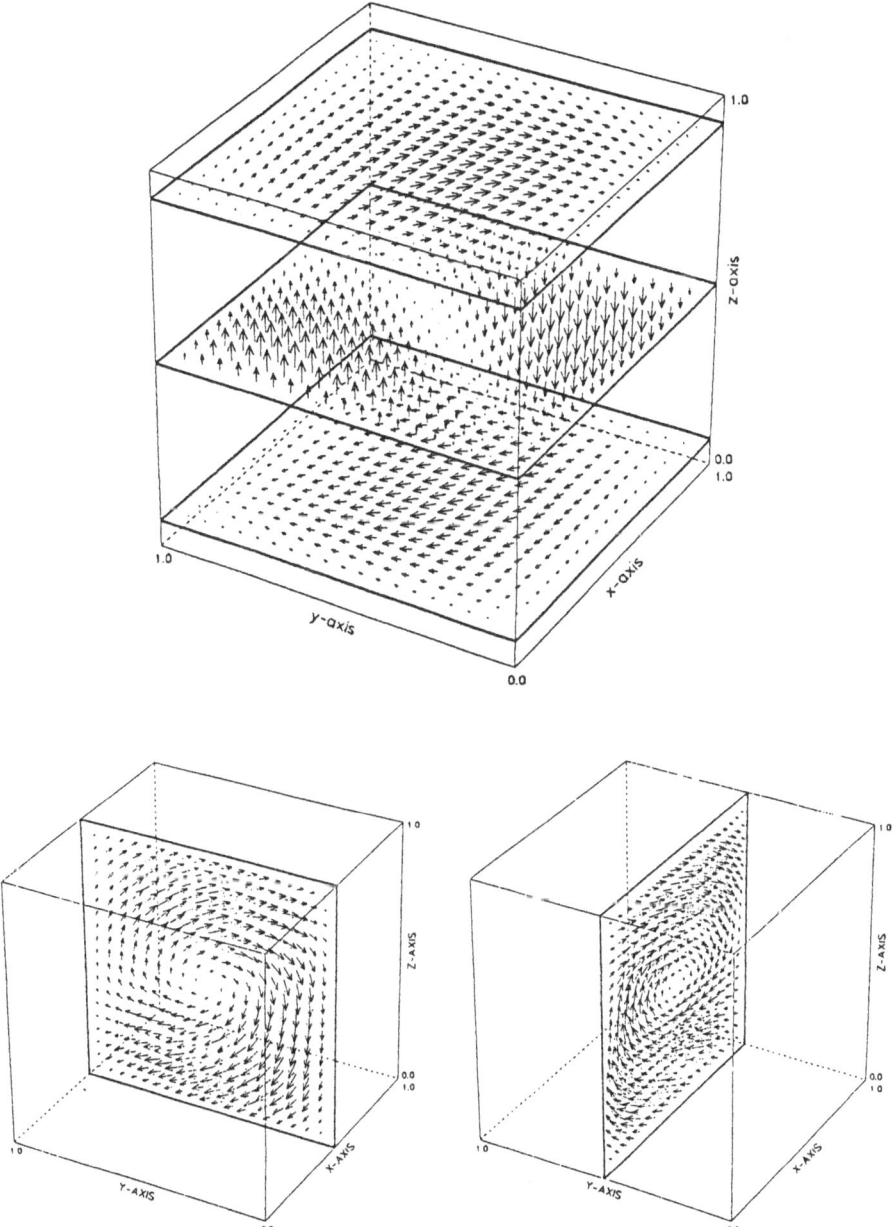

Fig. 5.5   Velocity vector plots of the solution of
           (5.7), (5.8) for different planes.

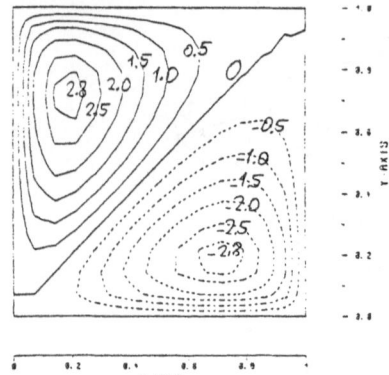 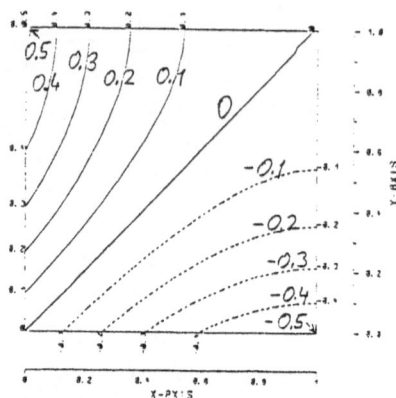

Fig. 5.6 Contour plot of
         the velocity com-
         ponent w in the
         mid-plane z=0.5.

Fig. 5.7 Contour plot of the
         temperature disturban-
         ce T in the mid-plane
         z=0.5.

software which can be executed and maintained on different types of vector computers (and similarly on general purpose computers). We think that we have found an acceptable compromise
between efficiency and robustness, even with the additional
requirement of a high degree of vectorisation. This is obtained only by an optimal data structure.

ACKNOWLEDGEMENT

     This research has been supported in part by the Deutsche
Forschungsgemeinschaft in the course of the FIDISOL project.

REFERENCES

[1] HOCKNEY,R.W., JESSHOPE,C.R.: "Parallel Computers", Adam
     Hilger Ltd, Bristol 1983.

[2] GENTZSCH,W.: "Vectorization of Computer Programs with
     Application to Computational Fluid Dynamics", Notes on
     Numerical Fluid Mechanics, vol.8, Friedr.Vieweg & Sohn,
     Braunschweig/Wiesbaden 1984.

[3] HOLLENBERG,J., "SX, NEC Corporation's supercomputer",

Supercomputer 1, Amsterdam Universities Computing Centre (SARA), May 1984, 6-8.

[4] ARNOLD,C.N., "Performance Evaluation of Three Automatic Vectorizer Packages", Proceedings of the 1982 Internat. Conf. on Parallel Processing, IEEE Computer Society Press, 235-242 (1982).

[5] KUCK,D.J., "Automatic Program Restructuring for High Speed Computation", CONPAR 81, Lecture Notes in Computer Science 111, Springer-Verlag, Berlin, New York 1981, 66-84.

[6] MIURA,K., UCHIDA,K., "FACOM Vector Processor VP-100/VP-200", High-Speed Computation, edited by J.S.Kowalik, Springer-Verlag, Berlin, New York, 127-138 (1984).

[7] Proposals Approved for FORTRAN 8X, X3J3 Committee, Document S8, January 1985.

[8] REID,J.K., "The Array Features in FORTRAN 8X", PDE-Software: Modules, Interfaces and Systems, edited by B.Engquist and T.Smedsaas, North-Holland, Amsterdam, New York, 351-354 (1984).

[9] REID,J.K., WILSON,A., "Array Processing in FORTRAN 8X", to appear in the Proceedings of the Conference "Vector and Parallel Processors in Computational Science II", Oxford, August 28-31, 1984, as a supplement to Computer Physics Communications (1985).

[10] WAGNER,J.L., "Status of Work Toward Revision of Programming Language Fortran", Signum Newsletter, vol.19, Nr.3, July 1984.

[11] DENNIS,J.B.,"Data Flow Supercomputers", Computer 13, 48-56 (1980).

[12] SCHÖNAUER,W., SCHNEPF,E., MÜLLER,H., "Designing PDE software for vector computers as a "data flow algorithm"", to appear in the Proceedings of the Conference "Vector and Parallel Processors in Computational Science II", Oxford, August 28-31, 1984, as a supplement to Computer Physics Communications (1985).

[13] SCHÖNAUER,W., MÜLLER,H., SCHNEPF,E., "Numerical tests with biconjugate gradient type methods", ZAMM 65, T391-393 (1984).

[14] VAN DER VORST,H.A., "Preconditioning by Incomplete Decompositions", ACCU-reeks 32, Academisch Computer Centrum Utrecht (1982).

[15] KIGHTLEY,J.R, JONES,I.P., "A comparison of conjugate gradient preconditionings for three-dimensional problems on a CRAY-1", CSS 162, Computer Science and Systems Division, AERE Harwell (1984).

[16] SCHÖNAUER,W., SCHNEPF,E., RAITH,K., "The redesign and vectorization of the SLDGL-program package for the self-adaptive solution of nonlinear systems of elliptic and parabolic PDE's", PDE Software: Modules, Interfaces and Systems, edited by B.Engquist and T.Smedsaas, North-Holland, Amsterdam (1984) pp. 41-66.

[17] SCHÖNAUER,W., SCHNEPF,E., MÜLLER,H., "PDE software for vector computers", Advances in Computer Methods for Partial Differential Equations - V, edited by R.Vichne-vetsky and R.S.Stepleman, IMACS 1984, pp. 258-267.

[18] SCHÖNAUER,W., SCHNEPF,E., MÜLLER,H., "Variable step size/variable order PDE solver with global optimisation", to appear in the proceedings of the 11th IMACS World Congress, Oslo, August 5-9, 1985.

[19] OERTEL,H.jun., "Thermische Zellularkonvektion", Habilitationsschrift Universität Karlsruhe (1979).

ADDENDUM: FIRST EXPERIENCES

ON THE FUJITSU VP 200

By the courtesy of the SIEMENS AG (we thank especially Dr. H.Gietl) we were able to test the VP 200 after the workshop. We think that these valuable experiences should be included as an addendum in these proceedings.

Table A1. Measured MFLOPS for simple operations.
(1): One Processor of CRAY X-MP, (2) CYBER 205 (2-pipe),
(3) VP 200. The vector length is n.

	$c_i = a_i + b_i$			$c_i = a_i * b_i$			$c_i = a_i / b_i$		
	(1)	(2)	(3)	(1)	(2)	(3)	(1)	(2)	(3)
n=10	8.7	8.7	6.4	8.7	8.7	6.4	6.6	4.8	3.7
100	41.0	47.2	47.3	39.7	49.0	47.3	18.7	12.9	20.8
1000	61.3	89.3	112.9	61.3	88.5	113.4	22.3	15.3	30.7
10000	67.5	98.0	152.9	67.0	97.1	152.4	23.3	15.6	31.7
	$c_i = a_i + s\, b_i$ (linked triad)			$g_i = (a_i+b_i)*c_i + (d_i-e_i)/f_i$			$s = s+a_i*b_i$ (scalar product)		
	(1)	(2)	(3)	(1)	(2)	(3)	(1)	(2)	(3)
n=10	14.7	10.4	12.5	27.2	8.0	16.2	5.7	6.5	9.1
100	80.3	77.6	91.9	71.6	31.8	93.0	34.8	38.5	77.9
1000	119.8	170.9	228.3	81.6	45.8	137.1	124.0	85.1	331.1
10000	138.0	194.2	307.7	85.2	46.6	142.6	163.8	97.1	361.0

In Table A1 there are presented measurements for three different vector computers including the VP 200. The bottleneck for the VP 200 are the two load/store pipes which can transfer only two words per cycle between main memory and V-registers. Thus instead of the peak performance of 267 MFLOPS for the simple add or mult only 2/3, i.e. 178 MFLOPS theoretical peak performance results because there would be needed 3 transfers/cycle. For n = 10 000 in Table A1 we find 152.9 resp. 152.4 MFLOPS. Similarly other operations may be limited by the load/store pipes. The most interesting quantity which we can deduce from Table A1 is Hockney's $n_{1/2}$. But there is immediately a difficulty: From the measurements for n = 10 or 100 for the add or mult we would get $n_{1/2} \simeq 500$ if we take the measured value for n = 10 000 as peak performance. But if we draw a <u>curve</u> through more dense measurements we get $n_{1/2} \simeq 250$ for the measured peak performance of 152.9 MFLOPS and we get $n_{1/2} \simeq 360$ if we take the theoretical peak performance of 178 MFLOPS. The discrepancy between these values results from the obvious fact that the "switching through" of the load/store pipes and the arithmetic pipes is a more complicated operation than the operation of a simple pipe. These measurements always are "from memory to memory". Nevertheless at any case $n_{1/2}$ has the highest value of the existing vector computers. The consequence is that the VP 200 will operate efficiently only for <u>long</u> vectors. Therefore the algorithms should be designed still more carefully for long and contiguous (to avoid bank conflicts) vectors than for the CYBER 205. Note that the $n_{1/2}$ for the VP 100 or the VP 400 (4 parallel pipes) is 0.5 or 2 times the $n_{1/2}$ of the VP 200.

Table A2. Measured MFLOPS for the multiplication of two n×n matrices by different algorithms. ①: <u>One</u> processor of CRAY X-MP, ② CYBER 205 (2-pipe), ③ VP 200.

	by scalar product			simultaneously by column			assembler or spec. algor.		
	①	②	③	①	②	③	① [1]	② [2]	③ [3]
n=10	5.5	4.9	11.2	11.8	6.9	21.1	93.7	12.4	
100	29.8	5.6	91.3	67.7	53.2	181.1	189.8	68.6	285
350	70.5	5.6	223.3	120.0	108.2	263.9	195.1	128.7	$\simeq$ 440

1) assembler coded, using an optimal chaining.
2) using 5-fold unrolling and special instructions.
3) using 2-fold unrolling (estimated for n = 350).

In Table A2 there are presented the measured MFLOPS for the multiplication of two n×n matrices by three different algorithms: The usual scalar product algorithm, the algorithm for the simultaneous computation of a whole column (gives contiguous vectors in Fortran) and for an assembler program or special algorithm. The memory-to-memory machine CYBER 205 cannot gain much efficiency by a special algorithm using some special instructions and unrolling of the outer loop. But the register-to-register machines can gain relatively much by programs which result in an optimal chaining and minimal transfer between main memory and V-registers (to keep operands as long as possible in the V-registers). Obviously also the Fujitsu compiler is not (yet?) able to do a really global chaining which should also include such provisions.

The measured timings for the solution of equation (5.1), corresponding to the values of Table 5.1 are for the VP 200: CPU scalar: 103.0 sec, CPU vector: 8.6 sec, thus gain g = 12.0. For the solution of equation (5.2) the timing which corresponds to the timings of (5.4) was 6.0 sec. Here 3-fold unrolling of the matrix-vector multiplication and all scratch files incore in COMMON has been used.

Here we note some supplementary remarks to section 2.4 which might be of general interest in judging the VP 200. The maximal length of the (reconfigurable) V-registers is 1024, therefore for vector length n > 1024 there is a (slightly) different timing. The vector pipes can start with the processing of the next vector as soon as the last element of the preceding vector has entered the pipe. Thus for such "chained" operations there is no start-up time for the following vectors and the pipes are continuously busy. We had no problems with the Fortran 77 compiler, all expected do-loops were vectorised. Some loops with suspicion of recursion had to be forced to vectorisation by compiler directives. The compiler does loop restructuring: A loop with vectorisable and non-vectorisable parts is split up to vectorise as many statements as possible. Nested loops might be interchanged to get an optimal execution. There is also a first step to a "global" chaining by keeping for n < 1024 in nested loops the operands as long as possible in the V-registers. But in spite of all nice special properties the price/performance relation and the availability of application software will decide about the success of the VP 200 on the vector computer market.

DEVELOPMENT OF EFFICIENT ALGORITHMS FOR THE SOLUTIONS OF FULL POTENTIAL AND EULER EQUATIONS ON VECTOR COMPUTERS

R. K. Jain, N. Kroll, R. Radespiel

DFVLR, Institute for Design Aerodynamics
3300 Braunschweig, Fed. Rep. Germany

SUMMARY

For the efficient use of vector computers, three problems in the area of computational aerodynamics are dealt with. These are: 3-D full potential equation, body-fitted grid generation for wing-body configurations and 3-D Euler equations. Special features of each of these problems are emphasized. Various techniques to overcome the bottlenecks appearing therein are described.

INTRODUCTION

With the advent of fast vector computers substantial progress has been achieved in computing three-dimensional transonic inviscid flows. The potential flow assumption forms the basis for a number of numerical methods, which have been developed during the last decade. In contrast, solutions of the Euler equations are expected to provide a more physical model for inviscid flows at the expense of increased numerical effort. Regardless of the type of model equations, generation of suitable body-fitted coordinate grids is one of the key features for an accurate numerical simulation.

To highlight the efficient use of vector computers, the present paper deals with special features of 3-D body-fitted grid generation, solutions of 3-D full potential equation and 3-D Euler equations. Using the potential flow model [1], vectorization of the numerical procedure, which is hampered by the mixed nature of the governing equations, is the main problem. Generation of body-fitted grids for wing-body configurations is based on the numerical solution of a system of Poisson equations using SLOR iteration scheme [2]. Using this approach problems both due to vectorization and I/O transfers arise, because an out-of-core memory storage is needed for very large grids. Explicit 3-D Euler equation methods are easily vectorized. However, there is an excessive demand on the CPU memory if a realistic aircraft configuration is to be analyzed. Therefore, a multi-block structured computer code has been developed [3]. Several aspects of this approach are discussed in greater detail.

FULL POTENTIAL EQUATION

The steady, inviscid motion of a compressible, isentropic fluid is described by the full potential flow equation. In non-conservation form using a cartesian coordinate system it is written as

$$(a^2 - \phi_x^2) \; \phi_{xx} + (a^2 - \phi_y^2) \; \phi_{yy} + (a^2 - \phi_z^2) \; \phi_{zz}$$

$$- 2 \; \phi_x \; \phi_y \; \phi_{xy} - 2 \; \phi_x \; \phi_z \; \phi_{xz} - 2 \; \phi_y \; \phi_z \; \phi_{yz} = 0. \qquad (1)$$

Here, ϕ is the velocity potential and a is the local speed of sound which can be determined from the energy equation

$$a^2 \;=\; a_o^2 - \frac{\gamma-1}{2} \; (\phi_x^2 + \phi_y^2 + \phi_z^2) \,, \qquad (2)$$

where a_o is the stagnation speed of sound and γ is the ratio of specific heats. When analysing transonic flow fields, the partial differential equation (1) is of mixed type:

- elliptic at subsonic points $\qquad (a^2 > \phi_x^2 + \phi_y^2 + \phi_z^2)$

- hyperbolic at supersonic points $(a^2 < \phi_x^2 + \phi_y^2 + \phi_z^2)$.

This mixed type nature of the governing equation, which requires a type dependent discretization scheme, is the main problem faced in the efficient vectorization of the solution process.

A well known numerical method for solving the 3-D full potential equation is the FLO 22 program developed by Jameson and Caughey [1]. Equation (1) written in a body-fitted coordinate system is discretized by a type dependent finite difference scheme using central differencing at subsonic points and upwind differencing at supersonic points. The resulting system of nonlinear difference equations is solved iteratively by a successive line over-relaxation in each spanwise section. The nonlinear equations are linearized using values of the previous iteration to calculate the first derivatives of the unknown ϕ. This iteration procedure leads first to setting up and then solving a tridiagonal system of linear equations for each horizontal line in each spanwise section and for each iteration. Setting up the systems of equations is the most time consuming part of the program and takes about 93 % of the total CPU-time. Since the computational procedure depends on the local flow type, it is bifurcated as sketched in fig. 1 and therefore, vectorization of this procedure in the computer code is hampered. The solution of the tridiagonal system by LU-decomposition is partially vectorized due to the recursive formulas occuring in this algorithm. But, there is no pressing need to implement more efficient algorithms as the solution of the tridiagonal system takes only 3 % of the total CPU-time.

Vectorization of the most time consuming part of the code can be achieved if the computations at subsonic and supersonic points are separated into different DO-loops. In the first loop, calculations common to all points are done. These are for the quantities, like metric terms and first order derivatives of ϕ, which do not depend on the local flow type. In the second loop, the tridiagonal system for a horizontal line is set up assuming that all grid points of that line are subsonic points. These two loops which take about 90 % of the total time, are now fully vectorized. The computational procedure continues calculating the local flow type at each grid point of the line. If there are no supersonic points, the tridiagonal system is solved, otherwise, the equations corresponding to supersonic points are modified. The loop over the supersonic points calculations is not vectorized, but it takes only about 3 % of the total CPU-time. Using this strategy, the code is fully vectorized for subsonic flows without additional computational work. For transonic flows, the most time consuming part is vectorized with some additional work at supersonic points. However, as the number of supersonic points is generally less than 5 % of the total number of grid points, the additional amount of work is negligible.

On CRAY computers, an alternative vectorization strategy is also possible. The CRAY-BNCHLIB routines CLUSFLT and CLUSFGE can be used to find clusters for subsonic and supersonic points. This leads to the separation of computations for subsonic and supersonic regions, both of which are now fully vectorized without additional work. However, the computer code using this strategy is not portable to other computers.

The original FLO 22 code and the vectorized version (FLO 22V) were used as benchmarks on various computers: two scaler machines, IBM 3081-K (DFVLR, FRG) and Siemens 7890-F (Siemens, FRG) and three vector machines, CRAY-1S (DFVLR, FRG), CRAY-XMP/1 (KFA,FRG) and Fujitsu VP-200 (Fujitsu, Japan). Since the vectorization was done without using any special CRAY-routines, identical versions of the FORTRAN code were run on all these machines. As a test case, transonic flow around the ONERA-M6 wing at $M_\infty = 0.84$ and $\alpha = 3.06^\circ$ was calculated using a 192 x 24 x 32 grid. The timings in CPU seconds are given in Table 1.

Table 1   Benchmarks with FLO 22 codes

Computer / Code	Scaler Computers		Vector Computers		
	IBM 3081-K	Siemens 7890-F	CRAY-1S	CRAY XMP/1	Fujitsu VP-200
FLO 22	3760	926	626	407	160
FLO 22V	-	965	202	139	75

The CPU of the Siemens 7890-F is nearly twice as fast as that of the IBM 3081-K, and furthermore, this machine has an additional functional unit to speed up the floating point operations. This leads to a CPU-time ratio of 4.1 in favor of the Siemens machine for the original FLO 22 code. On the Siemens machine, the additional amount of work due to vectorization is about 4 %. Vectorization on CRAY-1S, CRAY-XMP/1 and VP-200 gives a speed-up factor of about 3.1, 2.9 and 2.1, respectively. For the original FLO 22 code, the better performance of the VP-200 among the vector computers indicates the compiler of VP-200 vectorized significantly more DO-loops than the CFT 1.13 compiler of CRAY-XMP. The speed-up of CRAY-XMP over CRAY-1S is mainly due to the smaller clock period of the XMP. For the FLO 22 V code VP-200 machine is 1.85 times faster than CRAY-XMP/1. Main reasons for this speed-up for VP-200 are:

- faster vector unit
- higher throughput between vector registers and functional units

- larger capacity of vector registers (64 K bytes) and
- dynamically reconfigurable vector registers.

Vector registers of CRAY-XMP/1 are of fixed configuration and
have a capacity of 4 K bytes.
   Recently [4], the FLO 22V code has been run on one CPU of
the CRAY-XMP/48 at CRAY-Research, U.S.A., using an unreleased
compiler CFT 1.15. For this compiler, very long DO-loops were
split into smaller ones thereby achieving CPU-time of 78 seconds
for the test case.

BODY-FITTED GRID GENERATION IN THREE DIMENSIONS

   The second part of this paper deals with a method of grid
generation. Body-fitted grids around simple 3-D wings can be
generated by the use of a set of simple analytic funtions. This
was done in te previous example of FLO 22 using a square root
transformation along with shearing and stretching transformations.
Application of this approach to more complex configurations is
very much restricted. One of the alternatives in overcoming this
restriction is based on the work of Yu [2]. In this work, a 3-D
body-fitted grid about a wing-body configuration is generated by
solving a system of 3 coupled Poisson equations. In a compact
form, these equations in cartesian coordinates (x, y, z) are written as

$$\vec{\xi}_{xx} + \vec{\xi}_{yy} + \vec{\xi}_{zz} = \vec{P}\ (\xi, \eta, \zeta), \tag{3}$$

where $\vec{\xi} = [\xi,\ \eta,\ \zeta]^T$ and $\vec{P} = [P,\ Q,\ R]^T$. $\vec{\xi}$ are the computational coordinates and $\vec{P}$ are the source terms which control the
grid line spacings in the physical domain $(x,\ y,\ z)$. Solution of
equation (3) using equi-spaced net-work in the physical domain
to generate equi-spaced $\vec{\xi}$ lines involves complicated interpolation formulas. This difficulty is avoided when equation (3) is
transformed to the computational domain $\vec{\xi}$. By interchanging the
role of dependent and independent variables, the resulting system
of equations are written as follows:

$$A(\vec{x}_{\xi\xi} + \frac{J^2 P}{A}\ \vec{x}_\xi) + B\ (\vec{x}_{\eta\eta} + \frac{J^2 Q}{B}\ \vec{x}_\eta) + C\ (\vec{x}_{\zeta\zeta} + \frac{J^2 R}{C}\ \vec{x}_\zeta)$$

$$+ 2\ (D\ \vec{x}_{\xi\eta} + E\ \vec{x}_{\xi\zeta} + F\ \vec{x}_{\eta\zeta}) = 0, \tag{4}$$

where $\vec{x} = [x, y, z]^T$ and the Jacobian $J = \partial\vec{x}/\partial\vec{\xi}$. The coefficients
A to F are functions of the transformation coefficients $\vec{x}_{\vec{\xi}}$.
Positive or negative values of the underlined quantities in
equation (4) force $\vec{\xi}$-lines to be attracted to or repelled from
$\vec{\xi}_{max}$-lines, respectively. Values of $\vec{P}$ are derived along all inner
and outer boundaries and their values in the interior of the do-
main are interpolated. A wing-body combination with finite farfield boundaries in the physical domain is sketched in fig. 2(a)
and its counterpart in the computational domain is shown in
fig. 2(b).
   The system of equations (4) can be solved iteratively using
a successive line over-relaxation procedure. Using second order

38

accurate central differencing, the resulting system of difference equations are

$$-A \, \vec{\Delta}_{i-1,j,k}^{n+1} + \left[2A + \frac{2}{\omega} (B+C) \right] \vec{\Delta}_{i,j,k}^{n+1} - A \, \vec{\Delta}_{i+1,j,k}^{n+1} = \vec{R}_s^n + \vec{R}_H^{n+1}, \quad (5)$$

where $\vec{\Delta}^{n+1} = [(x^{n+1}-x^n), (y^{n+1}-y^n), (z^{n+1}-z^n)]^T$ is the correction vector, $\vec{R}_s^n$ is the residual of equation (4), $\vec{R}_H^{n+1}$ is the contribution of various terms on $(j-1)$-line and $(k-1)$-plane, ω is the relaxation parameter and the superscripts $n+1$ and n indicate present and the previous iteration number, respectively. The computational sequence in the numerical procedure to obtain the solution vector $\vec{x}$ is as follows:

```
  DO  N iterations
   DO  k = 1, KMAX
    DO  j = 2, JMAX-1
     DO  i = 2, IMAX-1
     Calculate A to F and P⃗
     DO  i = 2, IMAX-1
     Calculate Rs, RH for x
     DO  i = 2, IMAX-1                      fully vectorized
     Calculate Rs, RH for y
     DO  i = 2, IMAX-1
     Calculate Rs, RH for z
     Solution of tridiagonal system for Δ⃗  } Partially vectorized
    j, k, N end
```

The above algorithm consists of 4 fully vectorized inner DO-loops. In the first one, A to F and P, Q, R - related quantities are computed. The remaining three loops compute x-, y- and z-components of $\vec{R}_S$ and $\vec{R}_H$. This structure of the algorithm is well-suited for a multiprocessor computer where each component of $\vec{R}_S$ and $\vec{R}_H$ can be computed by each independent processor simultaneously. The tridiagonal system of equations (5) are solved for $\vec{\Delta}$ by LU-decomposition. This part of the algorithm is not fully vectorized, but this does not matter much as the amount of CPU-time taken here is a small fraction of the total CPU-time.

Using the above described procedure, a body-fitted grid around a wing-body configuration having a grid of 160 x 32 x 32 size was generated on CRAY-1S computer. The computations were done in vector mode with full in-core storage of $\vec{x}$-arrays. A sketch of surface grid net-work on a DFVLR wing-body model is shown in fig. 3(a) and details of the side-view of the grid at the wing-body junction are shown in fig. 3(b).

An out-of-core storage is resorted to in case of larger grids. These grids having about 300000 grid points are essential for complex configurations. These large grids are handled using multi-block structure approach. A grid of 192 x 32 x 32 size was split into two blocks of 192 x 32 x 16 and the solution vector $\vec{x}$ in each block is obtained for each iteration by successively buffering-out the computational results of each block. With this

2-block approach, an I/O transfer rate of 0.44 M Words/s on
CRAY-1S was achieved. For still larger grids which are needed in
the solution of Navier Stokes equations, multi-blocking can be
applied over j and k indices without disturbing the vectorized
loops over the i-index.

THREE DIMENSIONAL EULER EQUATIONS

The final part of this paper addresses the solution of
the 3D-Euler equations. The 3D time dependent Euler equations
form a set of 5 coupled partial differential equations of the
hyperbolic type. In the integral form these are written as

$$\frac{\partial}{\partial t} \iiint_{\Omega} \vec{W} d\Omega + \iint_{\partial \Omega} \bar{\bar{F}} \cdot \vec{n} \ ds = 0, \tag{6}$$

where Ω denotes a fixed region with boundary $\partial\Omega$ and outer normal $\vec{n}$. $\vec{W}$ and $\bar{\bar{F}}$ represent the vector of conserved quantities and
the corresponding flux tensor, respectively, and are given by

$$\vec{W} = \begin{bmatrix} \rho \\ \rho u \\ \rho v \\ \rho w \\ \rho E \end{bmatrix}, \bar{\bar{F}} = \begin{bmatrix} \rho u \ \vec{i}_x & + \rho v \ \vec{i}_y & + \rho w \ \vec{i}_z \\ (\rho u^2 + p) \vec{i}_x + \rho uv \ \vec{i}_y & + \rho uw \ \vec{i}_z \\ \rho uv \ i_x & + (\rho v^2 + p) \vec{i}_y + \rho vw \ \vec{i}_z \\ \rho uw \ \vec{i}_x & + \rho vw \ \vec{i}_y & + (\rho w^2 + p) \vec{i}_z \\ \rho uH \ \vec{i}_x & + \rho v H \ \vec{i}_y & + \rho w H \ \vec{i}_z \end{bmatrix},$$

where ρ, u, v, w, E and H are the density, cartesian velocity
components, total energy and total enthalpy, respectively. The
pressure p is obtained from total energy E using

$$p = \rho(\gamma - 1) \{E - \frac{1}{2} (u^2 + v^2 + w^2)\}.$$

The numerical method to solve equation (6) is based on the finite
volume scheme of Jameson et al.[5]. Semi-discretization of equation (6) separates spatial and time discretization. The finite
volume scheme is

$$\frac{d}{dt} (h_{i,j,k} \ \vec{W}_{i,j,k}) + \vec{Q}_{i,j,k} - \vec{D}_{i,j,k} = 0, \tag{7}$$

where $h_{i,j,k}$ denotes the volume of quadrilateral cells. $\vec{Q}_{i,j,k}$
represents the Euler flux balance for each cell. The fluxes at
the cell faces are calculated from averages of the values $\vec{W}$ assigned to the cell centers. To damp out high frequency oscillations and to capture shock waves without preshock oscillations,
a dissipative term $\vec{D}_{i,j,k}$ is added which is formed by a blend of
fourth and second differences. Details of the numerical procedure
are given in Radespiel and Kroll [3]. In our present code, Runge-
Kutta time stepping schemes are used. A 5-stage scheme with two
evaluations of dissipative terms has proved to be very efficient.
Several techniques to accelerate the convergence to the steady
state are applied. These are: local time stepping, enthalpy
damping, implicit residual averaging (allowing CFL number around
7) and successive grid refinement.

On the CRAY-1S vector computer, the inner DO-loops of the scheme are fully vectorized, yielding a rate of data processing (RDP) less than 0.5×10^{-4}, where RDP is defined as CPU-time in seconds required to update one elemental volume by one complete time step.

For efficient computations, about 30 variables for each elemental volume are stored. Discretization of a realistic aircraft configuration requires about 300000 elemental volumes, but the total memory of 1 M Word of CRAY-1 S limits the computational grid to about 20000 elemental volumes. Therefore, a multi-block structured computer code [3] has been developed. This approach has several advantages:

- The structure of the basic flow solver is simple and does not require any I/O transfer.

- All I/O transfers are done in special subroutines. Thus, it is easy to optimize data transfer procedures for any particular computer.

- The logic developed here to treat boundary conditions for each block allows a high flexibility of the code, so that it can be adapted to a variety of grid topologies with nearly unrestricted number of cells.

- The code can be extended to allow embedded regions, where time averaged Navier Stokes equations are solved.

On the otherhand, there are two problems connected with the multiblock approach, namely:

- The storage overhead for layers of dummy cells at block boundaries.

- Excessive I/O transfers to enable sufficient information exchanges between the blocks.

The storage overhead due to dummy variables can be limited to typically 15 % of the total block memory if a layer of only one cell thickness is used. Therefore, the treatment of fourth difference operator just outside the dummy cell is set equal to the corresponding difference operator just inside. This treatment of dissipative terms at inner cuts has proved to yield essentially the same results as obtained using unmodified dissipative terms.

The expense of CRAY-1S computations depend both on the amount of CPU-time and I/O waiting time in which data transfers between main memory and discs are executed. For a multi-block structured algorithm usually excessive I/O transfers are necessary to enable sufficient information exchanges between the blocks. Each change-over from one block to the next one requires about 0.5 M Words I/O transfers. All data transfers are done in special subroutines using very long records. Block field data are transferred using Buffer-IN/BUFFER-OUT simultaneously on both controllers of CRAY-1 S. Word addressable PUTWA/GETWA package is used for the exchange of block boundary data. The code, thus, achieves an I/O-transfer rate of about 55 MBIT/s. Nevertheless, the ratio between I/O and CPU-time is about 2.5 if a change-over from one block to the next occurs at each Runge-Kutta time step. This unacceptable high ratio can be reduced if several time steps are executed in each block before changing over to the next. To demonstrate the efficiency of this treatment, the decrease of averaged residuals for the transonic flow around ONERA-M 6 wing is discussed. The 96 x 20 x 20 coordinate grid has an O-type structure both in sectionwise and spanwise direction. The data about this grid was provided by Rizzi and is gratefully acknowledged. The grid has been divided into three blocks as illustrated in fig. 4. Fig. 5 (a) shows the rate of convergence to the steady state using averaged residuals for five cases: change-over from one block to the next at each 1st, 3rd, 5th, 8th and 12th time steps. Note, that the rate of convergence is nearly the same for the 8-step case as for the 1-step case. Obviously, the computational expense can be substantially reduced using this strategy. Fig. 5 (b) shows the rate of convergence against computational expense. The 5-step case yields a substantial reduction, the ratio between I/O and CPU-time being about 0.4. Fig. 6 displays spanwise variation of sectional lift for the converged solution. Two different block structures were arranged as indicated. For both cases, an identical lift distribution was obtained. Recently, a similar conclusion has been arrived at while computing transonic flow around DFVLR F4 wing. For these calculations, a coordinate grid of 224 x 32 x 40 cells has been used. The solution was obtained on three successive grids, the finest one being divided into 32 blocks as sketched in fig. 7. Even for this multi-block structured computational domain, no influence of block boundaries either on pressure distributions or on distributions of total pressure losses was observed.

CONCLUSIONS

Efficient vectorization of the FLO 22 code is achieved by calculating subsonic and supersonic regions separately. Impressive gains in CPU-time have been obtained. Body-fitted grid generation for a wing-body combination using a moderately sized grid has been done using completely in-core memory calculations. For larger grids requiring out-of-core memory storage, multi-block structure for the grid has been implemented. Excessive I/O penalties, while solving 3-D Euler equations requiring large grids, have been minimized using multi-block structured computational scheme achieving a value of 0.4 for the ratio of I/O time to CPU-time.

REFERENCES

[1] Jameson, A. and Caughey, D.A.: "Numerical calculation of transonic flow past a swept wing", ERDA Report COO-3077-140, New York Univ. (1977).

[2] Yu, N.J.: "Grid generation and transonic flow calculations for three-dimensional configurations", AIAA Paper No. 80-1391, (1980).

[3] Radespiel, R. and Kroll, N.: "Progress in the development of an efficient finite volume code for the three dimensional Euler equations", DFVLR FB 85-31 (1985).

[4] Cornelius, H.: Private communication, (1985).

[5] Jameson, A., Schmidt, W. and Turkel, E.: "Numerical solutions of the Euler equations by finite volume Methods using Runge-Kutta time-stepping schemes", AIAA Paper No. 81-1259 (1981).

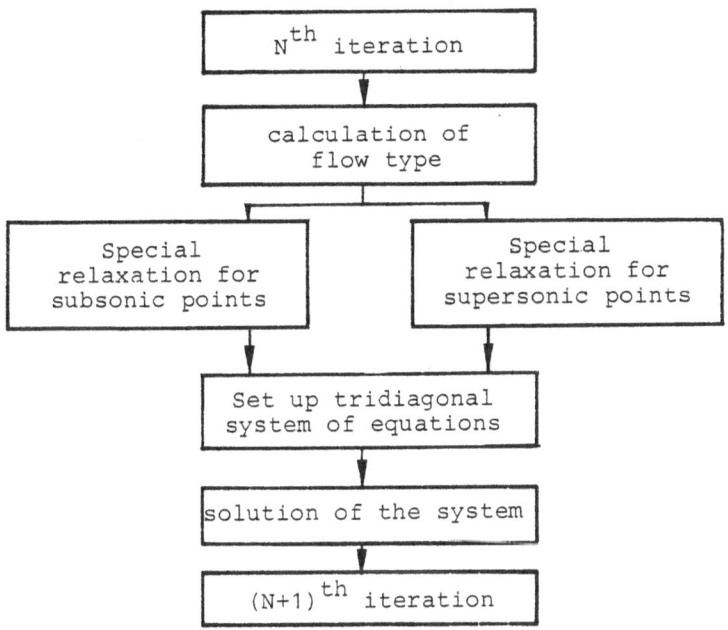

Fig. 1   Basic algorithm of the original FLO 22 code

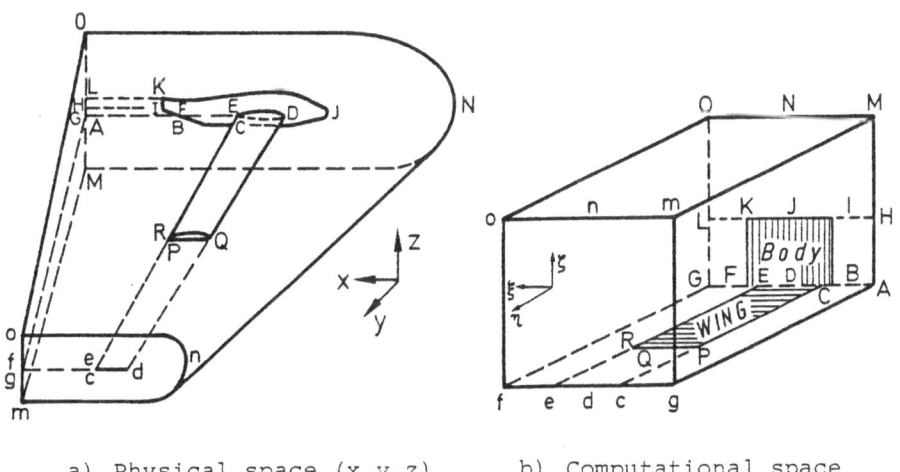

a) Physical space (x,y,z)

b) Computational space
   (ξ,η,ζ)

Fig. 2   Wing-body combination in physical
          and computational space

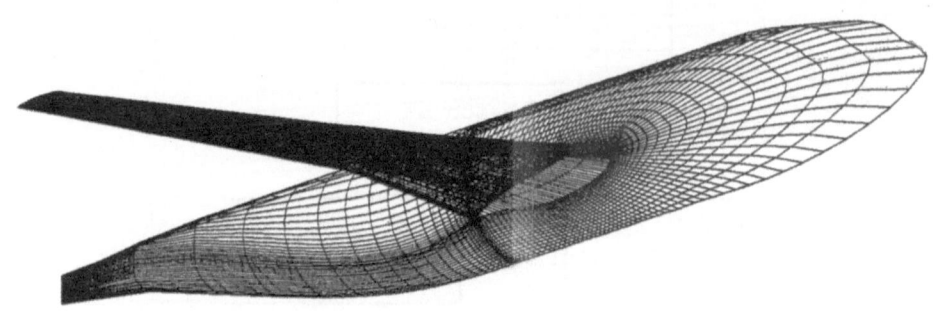

Fig. 3(a)   Surface grid network

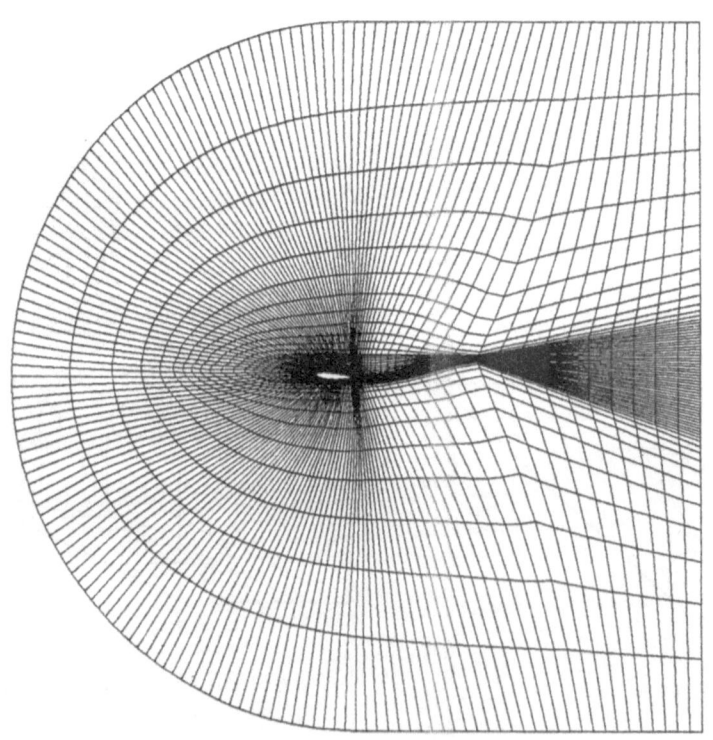

Fig. 3(b)   Grid network at wing-body junction

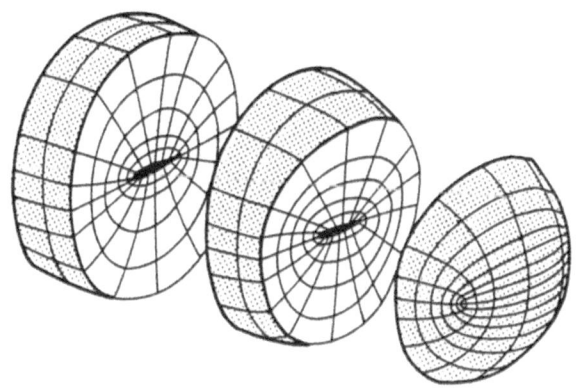

Fig. 4   O-O-coordinate grid divided into three blocks

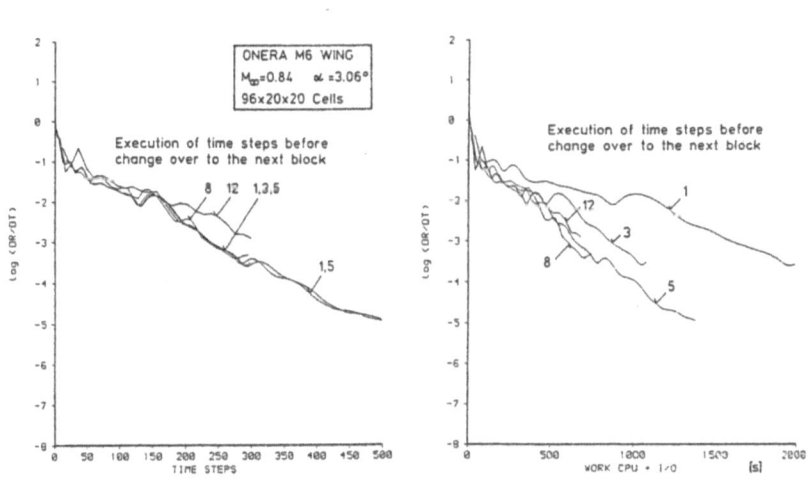

a) Averaged residuals over
   time steps

b) Averaged residuals over
   computational work

Fig. 5   Convergence behaviour and computational work for
         different block-loop versions

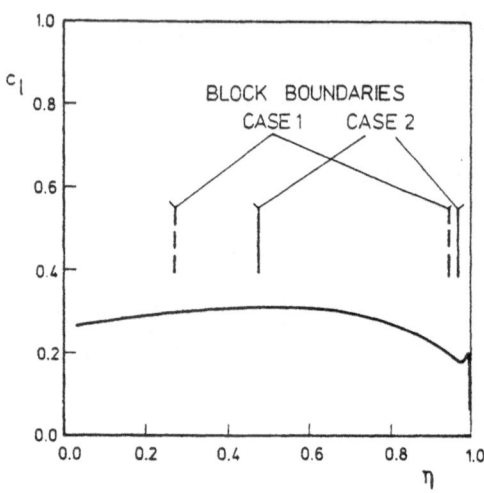

Fig. 6 Spanwise variation of sectional lift for
two different block structures,
ONERA M6 Wing, $M_\infty = 0.84$, $\alpha = 3.06^\circ$

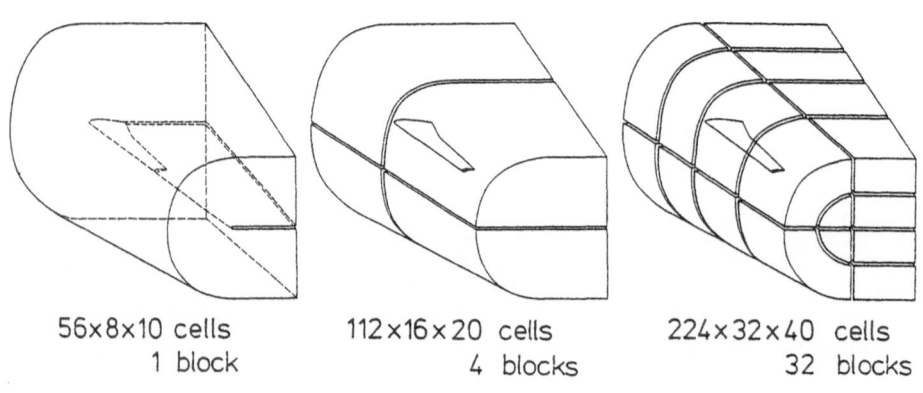

| 56×8×10 cells | 112×16×20 cells | 224×32×40 cells |
| 1 block | 4 blocks | 32 blocks |

Fig. 7 Block structured grids for wing or
wing-body configurations

IMPLEMENTATION OF 3D EXPLICIT EULER CODES

ON A CRAY-1S VECTOR COMPUTER

M. BORREL, J.L. MONTAGNE, M. NERON, J.P. VEUILLOT, A.M. VUILLOT

ONERA

B.P. 72 - 92322 CHATILLON CEDEX (FRANCE)

1 - INTRODUCTION -

The recent arrival at ONERA of a CRAY-1S vector computer, has raised the problem of adapting codes and of determining the most efficient choices in their features in order to fully exploit the vectorizing capabilities of the computer.

The methods and corresponding codes which are considered in this paper are based on explicit schemes, applied to the Euler equations, for 3D steady and unsteady problems. The flows may include surface discontinuities such as strong shocks or vortex sheets. These studies concern a code for steady flows with an isoenergetic assumption using a MacCormack type scheme, and a code for unsteady flows using a non centered scheme. Both codes are written for ijk structured grids.

The vectorization of an explicit algorithm on a structured grid is quite straightforward. Special attention has been brought to the understanding and improving of chaining. In order to point out some rules for programming the FORTRAN instructions, we have studied their translation into assembler language.

Beyond this aspect of CPU time minimization, computations performed on a 3D vectorized code have shown the important contribution of the I/Os to the total cost and their influence on the restitution time. Improvement of the I/Os and ressources management is considered in order to minimize both factors.

This result can be obtained by several means such as the reduction of the volume of I/Os, the introduction of synchronization between I/Os and computation, and various modifications of data management, which may interfere with the very structure of the code.

2 - DESCRIPTION OF THE CODES -

2.1 - The MacCormack type scheme Euler code (code A) :

a) General features of the method -

The computation of the 3D flow past a wing is performed by a pseudo-unsteady method of solution of the Euler equations with an iso-energetic assumption, the unsteady energy equation being replaced by the Bernoulli relation [1] [2] [3]. This replacement allows a reduction of the number of unknowns from 5 to 4. The unknowns are the density, and the momentum components. The continuity and momentum equations are integrated step by step using an explicit predictor-corrector scheme with a local time step technique. This scheme which is discretized directly in the physical space is identical to the MacCormack scheme in case of a regular rectangular mesh. A second order three points artificial viscosity term, and a fourth order five points smoothing term are added to the corrector in order to stabilize the solution. The structured grid, of C-O type, is composed of two symmetrical (I,J,K.) "boxes" corresponding to the upper and lower subdomains around a symmetrical wing.

b) 3-D Implementation -

The variables are stored by successive mesh surfaces in one grid direction (from the leading edge to the downstream boundary).

The five variables at each point are the density and the velocity components, which are modified at each iteration, and the local time step δt, which is usually reset every 100 iterations.

At each iteration, the computation of the predictor, of the corrector and of the dissipation terms are performed at the current point, each of these stages being successively executed for each of the mesh surfaces.

The computation of the predictor and corrector needs the knowledge of 2x9 geometrical coefficients, which are mesh derivatives. As this code has been written originally for a scalar computer, these derivatives are computed once at the beginning of the program, stored on disks, and read at each iteration.

In the basic version of the code, the minimum necessary storage represents 60 times the size of a grid surface (I = Cst).

c) Performances -

The inner loops in the subroutines performing computations at the current point have been vectorized, the scalar/vectorized time ratio being between 4. and 4.5. This value was obtained despite the short length of the loops (19 or 25) in this case of a 51x(2x25)x19= 43 350 point mesh. The maximum job size is 186 000 words. Calculation of the flowfield past a wing can be converged in 2 000 iterations which need 1 448s of CPU time (fig.1).

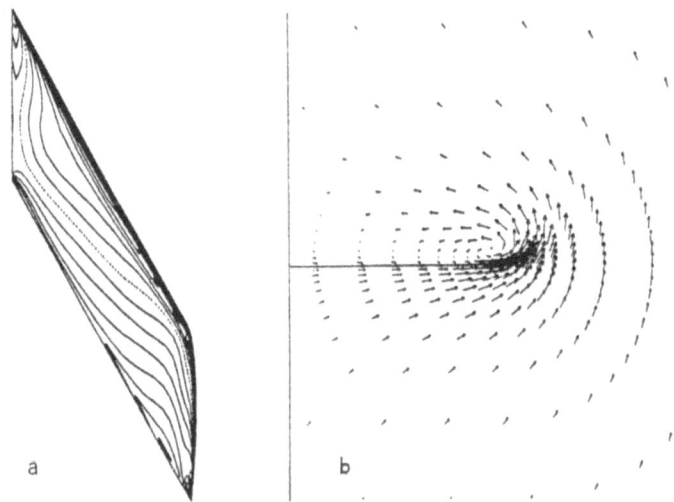

Fig.1 : Flow past the 60° swept angle AFV-D wing ; M_∞ = 0.92 ;
$\alpha = 4°$:
a) Isomach lines on the wing upper surface,
b) Velocity field on a transversal grid surface, at the trailing
edge.

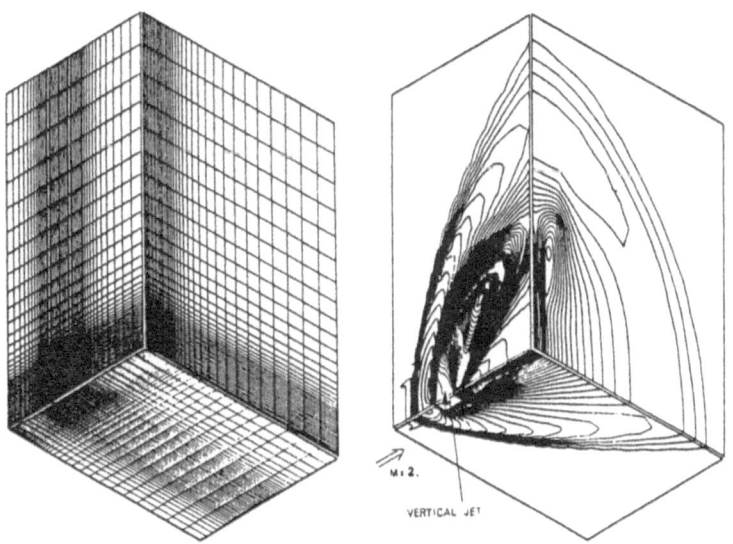

Fig.2 : Interacting jet with a supersonic flow. Grid and flow field on
the plane of symmetry, the horizontal plane and a transversal
plane.

The CPU time per iteration and per gridpoint is 16.7 μs. The CPU time represents 25% of the total cost, and the (I/O wait time)/(CPU time) ratio is about 3.

2.2 - The non-centered Euler Code (code B) -

a) General features of the method -
This code solves the unsteady Euler equations. It uses an explicit conservative finite volume scheme of second order of accuracy. This scheme is obtained from a first order Godunov type scheme by introducing appropriate corrections. The preliminary version which is presented uses a cartesian mesh. The vertices of the control volumes are centers of the cells.

The scheme is an extension of the two-dimensional version already developed at ONERA [4]. The calculation of the updated average values at time n+1 can be described in three stages :

First stage : at time n, a linear distribution of the variables is computed by interpolation from the average values in each control volume. For stability reasons, this interpolation takes into account a monotonicity constraint in each direction. This stage yields three slopes for the variables in each control volume.

Second stage : an increment of the variables over a half time step is evaluated by means of the slopes in each control volume.

Third stage : the updated average values of the conservative variables are obtained by computing fluxes crossing the interfaces between the control volumes at the half time step. The computation of fluxes uses a flux-vector splitting technique proposed by Van Leer [5] applied to the values at the center of the interfaces. These values are obtained from the linear distribution at time n, updated with the increment calculated at the second stage.

b) 3-D implementation -
The variables and the geometry are stored by successive planes in one grid direction. The calculation of a time step is performed plane per plane. For each plane this computation needs the storage of five variables (the density, the momentum components, and the energy) and of four parameters (the three coordinates x, y, z and the local time step) at each grid point of three planes. An auxiliary storage area with a size of 35 times the number of points in a plane is needed for temporary variables, among which are 15 slopes values per point in the plane. The minimum storage area corresponding to this implementation is 62 times the number of grid points in a plane. At each time step and for each plane successively, unblocked datasets are used, first for reading the four parameters and the five variables, then for writing five updated variables.

Finally, the boundary conditions are considered as corrections to the values found by the scheme on the external nodes. That leads to an additional storage, but it gives a great flexibility to the use of different types of boundary conditions.

50

c) Performances -

The loops in the subroutines performing computations at the current point have been vectorized by plane, but the treatment of boundary conditions have not yet been vectorized. The scalar/vectorized time ratio is 4..

For the test case of an interacting supersonic jet flow field (fig.2), a grid of 40x30x35 points has been used. This corresponds to 1 200 points in a plane. The maximal job size is 63 000 words. The steady state is reached within about 1 400 iterations and one hour of CPU time.

The CPU time per iteration and per grid point is 41 µs. The ratio of the I/O wait time to the CPU time is approximatly one.

3 - FORTRAN CODE OPTIMIZATION -

Classical vectorization of "DO" loops, such as, suppressing dependencies, switching inner and outer loops to have the longer index range for the inner loop, or splitting loops to lower the number of instructions in each one, yields satisfying performance improvements. However, it seems that better results could be obtained if one could choose, between several coding of loops, the one which allows the most efficient chaining and use of memory registers. Using the generated Cray Assembler Language (CAL) given by the directive CODE of the compiler, we have tried to find more precise rules for optimization of loops. Besides that, comparisons between single loop and double nested loop are shown.

The tested loop (fig.3) is taken from code B and calculates the slopes of each of the five variables RO, U, V, W and P. Each slope is an average of two differences involving sums of nine terms. The three dimensionnal arrays contain the values of the variables RO, ROU, ROV, ROW and ROE. The third index IP indicates the plane and the first one L indicates the position in the plane. This index is a combination of the first direction index I and the second direction index J. The second index K indicates the variable. In the present tests, the ranges of the indices are 49 for I, 2 for J, 100 for L, 5 for K and 3 for IP. The times given by the intrinsic function IRTC are in clock periods (CP). The gains are given with regard to the initial loop. The Fortran compiler is CFT 1.11. With its initial coding the loop lasts 139 403 CP. It cannot be vectorized by the compiler which has 2 560 words available for optimization.

This limitation suggests an easy modification to the loop, that consists in splitting it in several shorter loops. Here, splitting in three is necessary to obtain vectorization of each shorter loop. The advantage of this modification is that it is straightforward and purely local. But there are several drawbacks. At first auxiliary arrays for intermediate storage increase the main storage for the program. If single loops for index L are replaced by double nested loops for indices I and J, these arrays will not be too large. However, in this case the index I of the most inner loop has a shorter range than the initial index L. Finally, added instructions are necessary for the writing and reading of the auxiliary arrays. In fact one thing that makes the generated code so long is the complexity of index calculation with the three indices L, K and IP. Hence another way to have the code vectorized is to replace each

L-K-IP index array with 15 L-index arrays. 15 is the range of K multiplied by the range of IP. This modification is less straightforward than the first one but allows to keep a single loop with a long range index and does not require auxiliary arrays. The modified loop lasts 24527 CP and the gain is 5.68.

```
      AVE(A,B)=(A+B)*(EPS+A*B)/(A*A+B*B+2*EPS)
      DO 1000 ITER=1,30
      IT1=IRTC()
      DO 101 L=1,100
      RO1= 2.*U(L+1,1,IP  )+U(L+1+IMAX,1,IP  )+U(L+1-IMAX,1,IP  )
     1       +U(L+1,1,IPM1)+U(L+1+IMAX,1,IPM1)+U(L+1-IMAX,1,IPM1)
     2       +U(L+1,1,IPP1)+U(L+1+IMAX,1,IPP1)+U(L+1-IMAX,1,IPP1)
      QX1=(2.*U(L+1,2,IP  )+U(L+1+IMAX,2,IP  )+U(L+1-IMAX,2,IP  )
     1       +U(L+1,2,IPM1)+U(L+1+IMAX,2,IPM1)+U(L+1-IMAX,2,IPM1)
     2       +U(L+1,2,IPP1)+U(L+1+IMAX,2,IPP1)+U(L+1-IMAX,2,IPP1))/RO1
      QY1=(2.*U(L+1,3,IP  )+U(L+1+IMAX,3,IP  )+U(L+1-IMAX,3,IP  )
     1       +U(L+1,3,IPM1)+U(L+1+IMAX,3,IPM1)+U(L+1-IMAX,3,IPM1)
     2       +U(L+1,3,IPP1)+U(L+1+IMAX,3,IPP1)+U(L+1-IMAX,3,IPP1))/RO1
      QZ1=(2.*U(L+1,4,IP  )+U(L+1+IMAX,4,IP  )+U(L+1-IMAX,4,IP  )
     1       +U(L+1,4,IPM1)+U(L+1+IMAX,4,IPM1)+U(L+1-IMAX,4,IPM1)
     2       +U(L+1,4,IPP1)+U(L+1+IMAX,4,IPP1)+U(L+1-IMAX,4,IPP1))/RO1
      P1=(.1*GAM4)*(2.*U(L+1,5,IP)+U(L+1+IMAX,5,IP)+U(L+1-IMAX,5,IP)
     1          +U(L+1,5,IPM1)+U(L+1+IMAX,5,IPM1)+U(L+1-IMAX,5,IPM1)
     2          +U(L+1,5,IPP1)+U(L+1+IMAX,5,IPP1)+U(L+1-IMAX,5,IPP1)
     3          -.5*RO1*(QX1*QX1+QY1*QY1+QZ1*QZ1))
C
      RO2= 2.*U(L-1,1,IP  )+U(L-1+IMAX,1,IP  )+U(L-1-IMAX,1,IP  )
     1       +U(L-1,1,IPM1)+U(L-1+IMAX,1,IPM1)+U(L-1-IMAX,1,IPM1)
     2       +U(L-1,1,IPP1)+U(L-1+IMAX,1,IPP1)+U(L-1-IMAX,1,IPP1)
      QX2=(2.*U(L-1,2,IP  )+U(L-1+IMAX,2,IP  )+U(L-1-IMAX,2,IP  )
     1       +U(L-1,2,IPM1)+U(L-1+IMAX,2,IPM1)+U(L-1-IMAX,2,IPM1)
     2       +U(L-1,2,IPP1)+U(L-1+IMAX,2,IPP1)+U(L-1-IMAX,2,IPP1))/RO2
      QY2=(2.*U(L-1,3,IP  )+U(L-1+IMAX,3,IP  )+U(L-1-IMAX,3,IP  )
     1       +U(L-1,3,IPM1)+U(L-1+IMAX,3,IPM1)+U(L-1-IMAX,3,IPM1)
     2       +U(L-1,3,IPP1)+U(L-1+IMAX,3,IPP1)+U(L-1-IMAX,3,IPP1))/RO2
      QZ2=(2.*U(L-1,4,IP  )+U(L-1+IMAX,4,IP  )+U(L-1-IMAX,4,IP  )
     1       +U(L-1,4,IPM1)+U(L-1+IMAX,4,IPM1)+U(L-1-IMAX,4,IPM1)
     2       +U(L-1,4,IPP1)+U(L-1+IMAX,4,IPP1)+U(L-1-IMAX,4,IPP1))/RO2
      P2=(.1*GAM4)*(2.*U(L-1,5,IP)+U(L-1+IMAX,5,IP)+U(L-1-IMAX,5,IP)
     1          +U(L-1,5,IPM1)+U(L-1+IMAX,5,IPM1)+U(L-1-IMAX,5,IPM1)
     2          +U(L-1,5,IPP1)+U(L-1+IMAX,5,IPP1)+U(L-1-IMAX,5,IPP1)
     3          -.5*RO2*(QX2*QX2+QY2*QY2+QZ2*QZ2))
C
      ROIJ=U(L,1,IP)
      QXIJ=U(L,2,IP)/ROIJ
      QYIJ=U(L,3,IP)/ROIJ
      QZIJ=U(L,4,IP)/ROIJ
      PIJ=GAM4*(U(L,5,IP)-.5*ROIJ*(QXIJ**2+QYIJ**2+QZIJ**2))
C
      PTE(L,1,1)=AVE(.1*RO1-ROIJ,ROIJ-.1*RO2)
      PTE(L,2,1)=AVE(QX1-QXIJ,QXIJ-QX2)
      PTE(L,3,1)=AVE(QY1-QYIJ,QYIJ-QY2)
      PTE(L,4,1)=AVE(QZ1-QZIJ,QZIJ-QZ2)
      PTE(L,5,1)=AVE(P1 -PIJ ,PIJ -P2)
  101 CONTINUE
      IT2=IRTC()
      IT=IT2-IT1
      WRITE(6,*) 'CP= ',IT
 1000 CONTINUE
```

Fig.3 : Initial loop. CP = 139 403

Further improvements can be obtained by carefully examining the coding of each FORTRAN instruction. For instance each sum begins with an expression similar to 2UO(L+1)+UO(L+1+IMAX)+UO(L+1-IMAX). Figures 4 and 5 show how adding parentheses modify the generated code and the timing of the execution. Since the Cray 1-S has only one pointer for each vector register, a vector register can be used as an operand in only two

cases ; if it is totally free, i.e. containing the results of an already
completed instruction or if it is at a chaining point. That means that
it has just received its first element from another instruction. We see
on fig.4 that without parentheses (code B1) the evaluation is made from
left to right and the multiplication cannot chain with the addition.
When register V4 containing UO(L+1+IMAX) is at its chaining point,
register V5 containing 2UO(L+1) is not at its chaining point nor it
is free. The addition begins at time 141. With parentheses (code B2) the
expression is evaluated beginning from the most inner parentheses. Hence
VO containing UO(L+1+IMAX) is read at first, and when V7 containing
2UO(L+1) is at its chaining point, VO is free and chaining occurs. The
addition begins at time 86. But when it comes to read UO(L+1-IMAX) from
the memory, one can see that the difference between both timings
smoothes down because there is only one access to memory. Here the
functional unit memory is free at time 136. Since instructions must be
sequential and separated at least by one clock period, that makes time
142 for code B1 and time 136 for code B2. That illustrates the fact that
the more complex the FORTRAN instructions are, the less easy saving time
is by clever coding. For the overall loop adding parentheses to all
$2UO(L+1)+UO(L+1+IMAX)+UO(L+1-IMAX)$ type expressions changes the gain
from 5.68 to 5.72. Parentheses can also be added to the $QX1^2+QY1^2+QZ1^2$
type expressions. Replacing them by $(QX1^2+(QY1^2+QZ1^2))$ changes the gain
from 5.68 to 5.77. The final coding of the loop is given on fig.6. The
loop then lasts 24199 CP and the gain is 5.78.

FORTRAN	RO1 = 2. * UO (L+1) + UO (L+1+IMAX) + UO (L+1-IMAX)			
CAL	A6 B06 AO A7+A4 A7 A5+A2	AO A5+A4 A7 A5+A6 V4 ,AO,1 V3 V4+FV5	VL A3 S6 2. A1 A3−1 AO A7+A4	V6 ,AO,1 V5 S6*RV6 A2 B05 V2 ,AO,1

TIME (CP)	INSTRUCTION		FONCTIONAL UNIT FREE		RESULT CHAINING POINT		RESULT REGISTER FREE	OPERAND REGISTER FREE	
0 9	V6 V5	,AO,1 S6*RV6	M *	68 77	V6 V5	9 18	73 82	V6	73
68	V4	,AO,1	M	136	V4	77	141		
141 142	V3 V2	V4+FV5 ,AO,1	+ M	209 210	V3 V2	149 151	213 215	V4,V5	205

Fig.4 : Timing for code B1.

FORTRAN	RO1 = (2. * UO (L+1) + (UO (L+1+IMAX))) + UO (L+1-IMAX)				
CAL	VO ,AO,1	AO A5+A4	V1 AO,1	S6 2.	
	V7 S6*RV1	A1 A3-1	A2 BO5	A7 A5+A2	
	V6 V7+FVO	AO A7+A4	V5 ,AO,1	A7 PENTE5+3641	

TIME (CP)	INSTRUCTION		FONCTIONAL UNIT FREE	RESULT		OPERAND REGISTER FREE	
				CHAINING POINT	REGISTER FREE		
0	VO	,AO,1	M 68	VO 9	73		
68	V1	,AO,1	M 136	V1 77	141		
77	V7	S6*RV1	* 145	V7 86	150	V1 141	
86	V6	V7+FVO	+ 154	V6 94	158	V7,VO 150	
136	V5	,AO,1	M 204	V5 145	209		

Fig.5 : Timing for code B2.

The FORTRAN instructions of the same loop have been taken to compare the efficiency of programming with a single loop or double nested loops. For this purpose, the same instructions have been written in three ways : a first time with one vectorized loop over index L=I+(J-1)*IMAX, including fictitious nodes at boundaries, a second version with two nested loops over the indexes I and J, the inner one over index I beeing vectorized, and a third one like the second version with the inner loop not vectorized. The CPU time needed to execute these loops has been compared for different values of I and J. The following levels of vectorization (LV) for the inner loop have been found :

I	10	20	50	100	200
LV	2.5	3.6	5.1	5.5	5.7

```
       AVE(A,B)=(A+B)*(A*B+EPS)/(A*A+B*B+2*EPS)
       DO 1000 ITER=1,30
       IT1=IRTC()
CDIR$ CODE
       DO 101 L=1,100
       RO1= (2.*UO (L+1)+(UO (L+1+IMAX)))+UO (L+1-IMAX)
     1      +UOP(L+1)+ UOP(L+1+IMAX)  +UOP(L+1-IMAX)
     2      +UOM(L+1)+ UOM(L+1+IMAX)  +UOM(L+1-IMAX)
       QX1=((2.*U1 (L+1)+(U1 (L+1+IMAX)))+U1 (L+1-IMAX)
     1      +U1P(L+1)+ U1P(L+1+IMAX)  +U1P(L+1-IMAX)
     2      +U1M(L+1)+ U1M(L+1+IMAX)  +U1M(L+1-IMAX))/RO1
       QY1=((2.*U2 (L+1)+(U2 (L+1+IMAX)))+U2 (L+1-IMAX)
     1      +U2P(L+1)+ U2P(L+1+IMAX)  +U2P(L+1-IMAX)
     2      +U2M(L+1)+ U2M(L+1+IMAX)  +U2M(L+1-IMAX))/RO1
       QZ1=((2.*U3 (L+1)+(U3 (L+1+IMAX)))+U3 (L+1-IMAX)
     1      +U3P(L+1)+ U3P(L+1+IMAX)  +U3P(L+1-IMAX)
     2      +U3M(L+1)+ U3M(L+1+IMAX)  +U3M(L+1-IMAX))/RO1
       P1=(.1*GAM4)*((2.*U4 (L+1)+(U4 (L+1+IMAX)))+U4 (L+1-IMAX)
     1               +U4P(L+1)+ U4P(L+1+IMAX)  +U4P(L+1-IMAX)
     2               +U4M(L+1)+ U4M(L+1+IMAX)  +U4M(L+1-IMAX)
     3               -.5*RO1*(QX1*QX1+(QY1*QY1+QZ1*QZ1)))
C
       RO2= (2.*UO (L-1)+(UO (L-1+IMAX)))+UO (L-1-IMAX)
     1      +UOP(L-1)+ UOP(L-1+IMAX)  +UOP(L-1-IMAX)
     2      +UOM(L-1)+ UOM(L-1+IMAX)  +UOM(L-1-IMAX)
       QX2=((2.*U1 (L-1)+(U1 (L-1+IMAX)))+U1 (L-1-IMAX)
     1      +U1P(L-1)+ U1P(L-1+IMAX)  +U1P(L-1-IMAX)
     2      +U1M(L-1)+ U1M(L-1+IMAX)  +U1M(L-1-IMAX))/RO2
       QY2=((2.*U2 (L-1)+(U2 (L-1+IMAX)))+U2 (L-1-IMAX)
     1      +U2P(L-1)+ U2P(L-1+IMAX)  +U2P(L-1-IMAX)
     2      +U2M(L-1)+ U2M(L-1+IMAX)  +U2M(L-1-IMAX))/RO2
       QZ2=((2.*U3 (L-1)+(U3 (L-1+IMAX)))+U3 (L-1-IMAX)
     1      +U3P(L-1)+ U3P(L-1+IMAX)  +U3P(L-1-IMAX)
     2      +U3M(L-1)+ U3M(L-1+IMAX)  +U3M(L-1-IMAX))/RO2
       P2=(.1*GAM4)*((2.*U4 (L-1)+(U4 (L-1+IMAX)))+U4 (L-1-IMAX)
     1               +U4P(L-1)+ U4P(L-1+IMAX)  +U4P(L-1-IMAX)
     2               +U4M(L-1)+ U4M(L-1+IMAX)  +U4M(L-1-IMAX)
     3               -.5*RO2*(QX2*QX2+(QY2*QY2+QZ2*QZ2)))
C
       ROIJ=UO(L)
       QXIJ=U1(L)/ROIJ
       QYIJ=U2(L)/ROIJ
       QZIJ=U3(L)/ROIJ
       PIJ=GAM4*(U4(L)-.5*ROIJ*(QXIJ**2+(QYIJ**2+QZIJ**2)))
C
       WO(L)=AVE(.1*RO1-ROIJ,ROIJ-.1*RO2)
       W1(L)=AVE(QX1-QXIJ,QXIJ-QX2)
       W2(L)=AVE(QY1-QYIJ,QYIJ-QY2)
       W3(L)=AVE(QZ1-QZIJ,QZIJ-QZ2)
       W4(L)=AVE(P1-PIJ,PIJ-P2)
  101 CONTINUE
C
CDIR$ NOCODE
       IT2=IRTC()
       IT=IT2-IT1
       WRITE(6,*) 'CP= ',IT
 1000 CONTINUE
```

Fig.6 : Optimized loop. CP = 24 119

The relative gains of the single loop over the double nested loops are reported in per cent on this chart :

I / J	10	20	50	100	200
2	13.3	7.9	1.5	1.0	1.1
3	17.3	10.0	2.0	1.6	1.1
5	20.3	10.2	2.2	1.8	1.3
10	21.0	10.3	2.7	2.0	1.4
20	21.5	10.7	2.9	2.2	1.5
50	22.0	10.8	3.1	2.1	1.4

From these values, it appears that, though more arithmetic operations are done with the single loop due to the fictitious nodes introduced, the gain is always positive. However, with grids of realistic dimensions ($I > 50$), this gain becomes negligible.

An initiation to the use of the generated CAL may be useful to a FORTRAN programmer in order to better understand the reasons of the performance of his coding. However a systematic use of this tool seems useless. When the loops are complex and once some rules depending on the compiler and possibilities of the computer are respected, the compiler most often finds the best optimization solution. Some rules for the CRAY 1-S, compiler CFT 1.11 are :

- prefer one-dimensional to multidimensional arrays in order to shorten the generated code
- avoid use of temporary variables in order to suppress wait time due to the saving of the value of these variables, which stops chaining.
- knowing the precedence for evaluation of arithmetical expressions, organize them to enhance chaining and overlapping of instructions.
- order statements so as to free the registers and save memory access.

Furthermore, the tests on single and double nested loops showed that the negligible benefits yielded by the single loops do not justify the effort of implementation.

Though the examples involve relatively coarse grids, the presented codes are intended to be used on domains up to one million grid points. Four factors have been considered for efficiency of the Input/Output transfer — the intrinsic speed of transfer — the amount of information — the data management — the synchronization. The conclusions given in this section are based upon some improvements tested on code A and systematic tests described in the next section for the multidomain strategy developed on code B. It should be noticed that the results have been obtained in the environment of a computer shared by several users. Tests have been performed several times in order to estimate the variations induced by this fluctuating environment.

a) Speed of transfers —

No special device was available on our CRAY-1S for accelerating the transfers, apart from the use of multidisk access (four files can be accessed simultaneously). We propose a comparison between the time of transfer for five variables, and the CPU time for computation of a time step on a plane. The transfer is made using unblocked datasets and asynchronous "buffer in" "buffer out" orders. For the tests on the multidomain strategy with code B different sizes of records were considered and the transfers performed several thousands of times. The number of clock periods has been counted using the function IRTC for both processes, and the time has been estimated as the minimum value of this number. These times are plotted on figure 7 as function of the size of the record. The plot for the MacCormack scheme CPU time is estimated from a ratio of 2.1 with respect to the non-centered scheme. These results show that, when no external perturbation slows the transfers down, the time is proportional to the size of the record. The time of transfer is approximatly 3 times shorter than the computation with code B, but this ratio is only 1.3 with code A. The use of four disk access, in code B, only saves 30% of the time of transfer, but the number of tests has been small (10).

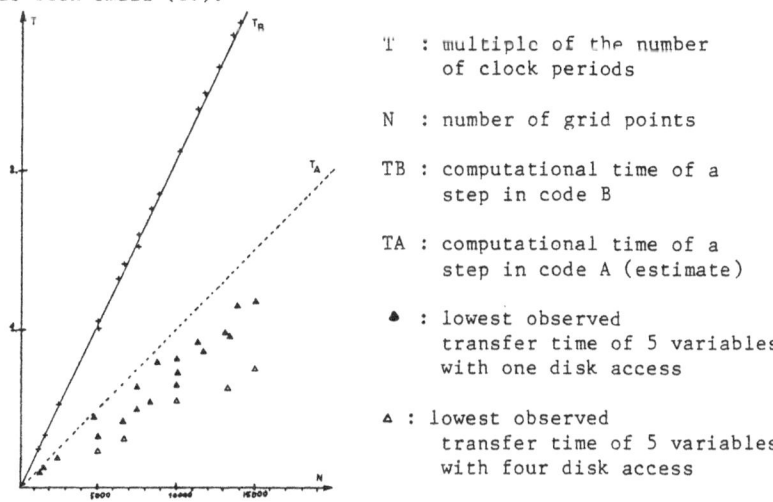

T : multiple of the number of clock periods

N : number of grid points

TB : computational time of a step in code B

TA : computational time of a step in code A (estimate)

▲ : lowest observed transfer time of 5 variables with one disk access

△ : lowest observed transfer time of 5 variables with four disk access

Fig. 7 : Time as a function of the size ·N.

b) Amount of information –

For an explicit code, the amount of information is fixed by the number of variables and of geometrical coefficients. In code B, at each iteration, 9 values per grid point are read (5 for the variables, 4 for the geometry), and 5 are written (the new values of the variables). In code A, 23 values per grid point are read at each iteration (5 for the variables, 18 for the geometrical coefficients). On the scalar computer on which this code was developed, this way of proceeding saved CPU time. On the CRAY 1S vector computer, for code A, and in the particular case of the accounting method used, the CPU time cost represents 26% of the total cost, the product "memory by CPU time", 2%, the product "Memory by I/O wait time", 17%, and the "Disk sectors moved", 55%.

This last item represents the total volume of I/O. If nothing was done to reduce this volume, an optimisation could only lower the 17% corresponding to the I/O wait time. Recomputing the geometrical derivatives at each iteration would permit to replace the corresponding 18 values per grid point in the storage by the only 3 values of the coordinates. This would bring reductions of the I/O wait time and of the in core memory, as these geometrical derivatives, which would be computed only when needed at the current point, would require no special array. Moreover, the "Disk sector moved", lowered by the ratio 13/28 would decrease to 25.5% of the present total cost, the benefit being of 29,5%.

Although this modification, which needs reprogramming a large part of the code hasn't been brought here, this aspect seems of major importance for the developing of any new code.

c) Resources management –

The total volume of I/Os being fixed, accounted I/O wait time can depend on the number of calls for I/Os. This is specially true in a multi-users environment. A first rule for lowering the calls for I/Os consists in optimizing each I/O order f.e. by using unblocked datasets, or by choosing for the buffers a size not smaller than the record size. Another rule consists in gathering the variables which are transferred at each iteration. For example, a comparison has been made on code A between the basic version, in which the 9 predictor and the 9 corrector geometrical coefficients were read separately, and a new version in which they were read altogether. Although the results appeared to be very sensitive to the environment on the computer, the (I/O wait time)/(CPU time) ratio was about 8, in the first version, and 6 in the second one. This can be confirmed by counting the number of clock periods. There is not proportionality between the time for I/O and the size of the record in a usual multi-users environment. It may be verified only when few users are executing simultaneously.

An important parameter to optimize is the length of the buffers used for the transfer of the solution in the main storage, which may correspond to a plane, or a block or a part of it. This choice is quite free in an explicit method, if no access problem is present, unlike in an implicit method.

It must be noted that gains on the I/O wait time, with corresponding increases on the in-core memory may, in some cases, result in low savings on the total cost, if not negative. Such a situation, however may be prefered in case of a high number of mesh points, even if the "waiting time for I/O" is not accounted, in order to keep reasonable restitution times.

d) Synchronization –

Synchronization is an efficient way to reduce the I/O wait time. First improvements of code A have led to a decrease by 75% the product (memory)x(I/O wait time).

In the basic versions of codes A and B, I/O orders, as usual, are performed at each iteration, with not enough time before the use of the transfered data to permit a good synchronization. Plots of fig.7 show that the time of transfer/time of computation ratio is about .25 for code B and somewhat below 1. in the present version of code A. Such a ratio, lower than 1. allows complete synchronization as the transfers can be done during the computation. Synchronization needs an increase of the in-core memory in order to store simultaneously the unknowns on two or more grid surfaces. For grid surface sizes inferior to 10 000 points, the available memory of the CRAY 1S is enough. For larger dimensions, a multi-domain technique such as exposed in [5] is necessary in order to permit synchronization.

5 – MULTIDOMAIN STRATEGY –

a) Description –

In case of a large size for the planes of records, the governing idea is to split the domain in NSD slices in a direction perpendicular to the progression. (See fig.8 for diagram and notations). The smaller amount of points in a subdomain allows to gather the records by blocks of planes. This implementation is considered here for code B. The computation progresses in the z direction, and the domain is splitted in the y direction in slices of equal numbers of points. The subdomains are successively computed on a time step. They are matched using a technique proposed in [6] . This induces a minimum of additional storage because an overlapping on only one plane xz is necessary for matching the subdomains. In each subdomain, the same algorithm is used as in the basic version, but two blocks of NPB planes of 5 variables and 4 coordinates are stored in core. The minimum value of NPB for the progression is 2, and a synchronization between the I/O transfers and the computations begins when NPB=3.

As code B needs 35 auxiliary variables,the size of the storage area is equal to :

$$TMC = NPTP/NSD * (35 + 2*9*NPB) + 5*NX*NY .$$

The last term is the area needed for matching the subdomains. Tests have been performed with uniform flow field, for five different sizes of the domain (fig.9), the values of the parameters NPB and NSD varying in the range 1 to 4 for NSD, 1 to 12 for NPB. The value 1 for NPB corresponds to the basic version of data managment by plane.

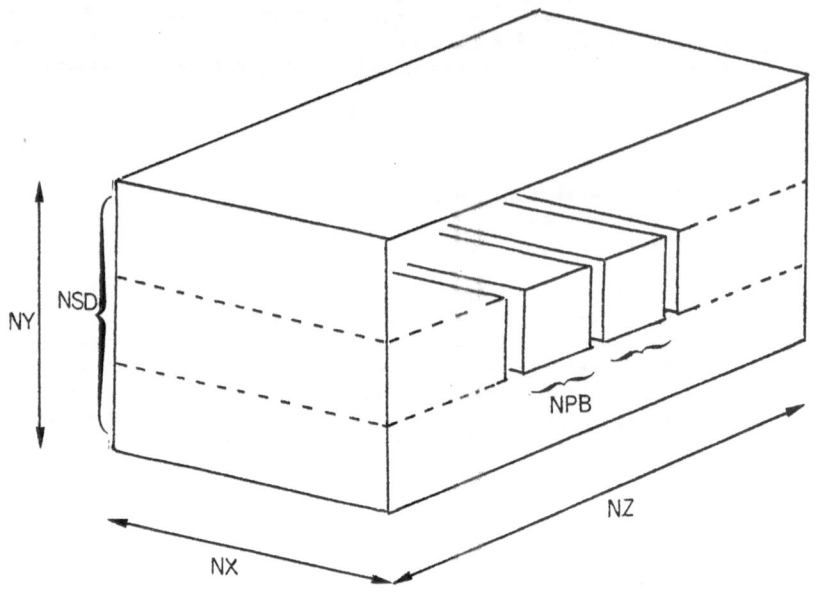

```
NSD       = number of subdomains
NPTP      = total number of points in the plane xy
NPB       = number of planes in a block
NX,NY,NZ  = number of points in the directions x,y,z
```

Fig.8 : Multidomain stategy notations.

case	total number of points	NX	NY	NZ	NPTP	maximum job size in the basic version (words)
A	30 000	25	40	30	1 000	121 000
B	200 000	50	40	100	2 000	287 000
C	500 000	50	100	100	5 000	470 000
D	800 000	80	100	100	8 000	697 000
E	1 000 000	100	100	100	10 000	830 000

Fig.9 : Description of the five test cases. NPTP : size of the plane
XY ; NX, NY, NZ : numbers of points in x,y,z directions.

b) Data management -

The size of the storage TMC has to be compared to the size
TMC$\emptyset$=62$\ast$NPTP obtained for the basic version. Due to the relatively
high number (35) of auxiliary variables, the reduction of the size of
the plane has a favourable effect on the value NPB of planes per block
if we assume that TMC is equal to TMC$\emptyset$.

It has been noticed previously that the I/O wait time is sensitive to
the number of calls for I/O transfers. For a given case, this number of
calls for I/O over a time step is reduced by the factor NPB/NSD with
regard to the basic version. The maximum values of NPB and of NPB/NSD
are given below for a constant storage TMC=TMC0, as functions of the
number of subdomains.

NSD	1	2	3	4	5	∞
NPB	1	4	8	11	15	—
NPB/NSD	1	2	2.67	2.75	3.	3.44

The factor NPB/NSD has a limited asymptotic value, but a value of 3 is
obtained for a small NSD. Unfortunatly, the effect is basically
dependent on the load of the computer, and the large discrepancy in the
results does not allow to show a real correlation between the I/O wait
time and the ratio NPB/NSD. On the other hand, figure 10 shows the (I/O
wait time)/(CPU time) ratio as a function of the number of points in a
subdomain, and of NPB, the number of planes gathered in a block. Again,
a large discrepancy appears, but there is a tendancy for the I/O wait
time to diminish as NPB increases. The figure also shows bad results
when the size of the plane is too small. This is an indication for not
using too many small subdomains. Very low I/O wait time occuring for
values of NPB upper than 5 are due to synchronization.

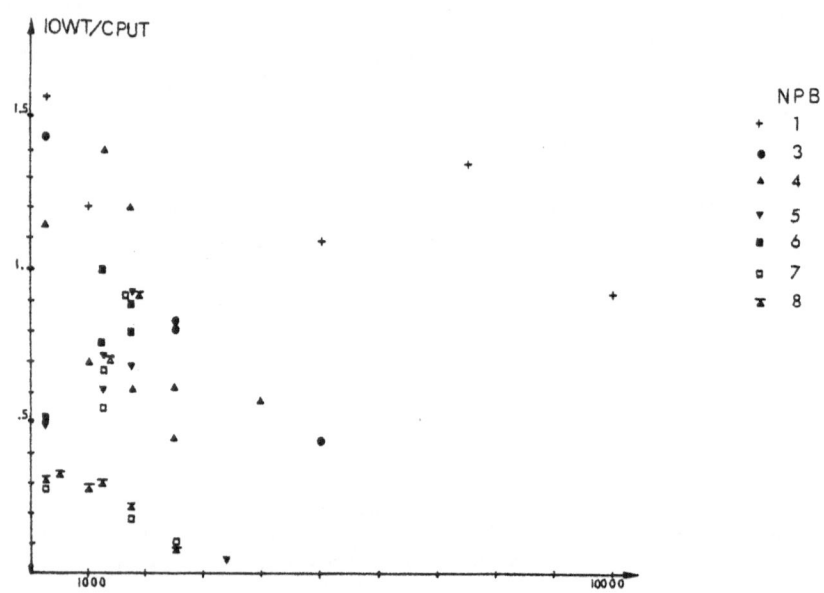

Fig.10 : (I/O wait time)/(CPU time) ratio versus the number of grid
points of the plane xy in a subdomain, for different values
of NPB.

c) Synchronization -

For the I/O transfer, the whole computation of a time step is considered
as a unit. During the current process, the computation is performed on
an area of NPB plans, the other area is used for storing either the
updated variables in the planes of the previous block, or the variables
at time n in the next block. Because writing and reading transfers
correspond to the same area in core, one must wait the end of the other
for beginning. Furthermore, the writing must not be started before the
end of the calculation of the first plane in a block, and the reading
must end before the beginning of the computation of the last plane in
the same block. The implementation choices for the tests lead to the
fact that the transfers of NPB planes can be done during the computation
of a number of planes reported below as a function of NPB :

NPB	1	3	4	5	6	7	8	9	10	11	12	13
Reading	0	0	0	1	2	3	3	4	4	5	5	6
Writing	0	1	2	2	2	3	3	4	4	5	5	6

For values of NPB larger than 9, the transfers must be twice faster than the computations in order to achieve synchronization. The observed ratio of 3.0 for code B (fig.7), allows synchronization with a value of 6 (even 5 with multidisk access) for NPB. This explains the occurrence of some low (I/O wait time)/(CPU time) ratios in fig.10.

6 - CONCLUSION -

Adapting existing codes, or developing new ones on a vector computer such as the CRAY 1S needs special attention from the user concerning both vectorization and ressources management, in order to get the best of the machine.

Apart from the application of a few simple rules, which can lead rapidly to some results, the choices to be made, which may influence the very structure of the codes, have to be watched carefully, because of the amount of work implied by latter modifications. In most cases however, finer optimization needs numerous tries, in a somewhat experimental way, when no more rule can be applied, either because of a lack of information, or because of the specificity of the case.

As for vectorization and chaining, some programming rules can be edicted for writing loops in FORTRAN. Further gains may be obtained by the study of the translation of the code into assembler language, or even further, by the use of the assembler language itself for some often called routines. It can be noted that some bottlenecks such as the presence of a unique pointer for the registers or of a unique operator for memory transfers have been suppressed on new versions of CRAY computers.

Concerning the ressources management, and unlike in a scalar computer, one of the main points consists in the necessity to minimize the I/Os, even if it needs computing again some coefficients at every iteration, as the I/Os may represent a major part of the computational cost. This rule is especially important for the development of a new code.

Apart from the volume itself of the I/Os, the "waiting time for I/Os" can be lowered, for example by enlarging buffer sizes, and, mainly, by improving synchronization, despite increases of the in-core memory. This may lead to complex optimizations, in order not only to minimize the global cost, but also for example to find a compromise between this cost and the restitution time, especially in case of large mesh sizes.

In the last cases of large mesh sizes, synchronization may need special subdomain techniques in order to be complete, the corresponding parts of neighbouring grid surfaces being read together in advance. This technique, although leading to more complex codes may save important costs for large mesh sizes.

In all aspects, the best choices to be made may depend on local or time dependent parameters such as the capabilities of the software, the configuration of the computer, the environment of the job at the time it is run, and also the accounting method of the whole calculation.

Aknowlegments : The authors would like to thank H. BOILLOT (ONERA Informatics department) for his valuable help about coding on the CRAY-1.

REFERENCES

[1] H.VIVIAND, J.P. VEUILLOT

"Méthodes pseudo-instationnaires pour le calcul d'écoulements transsoniques."
ONERA Publication n°1978-4 (English translation : ESA-TT 561).

[2] C. KOECK, M. NERON

"Computations of Three-Dimensional Transonic Inviscid Flows on a Wing by Pseudo-Unsteady Resolution of the Euler Equations".
Esd. M. PANDOLFI, R. PIVA, Proceedings 5th GAMM Conf. Num. Meth., VIEWEG VERLAG, 1984.

[3] F. MANIE, M. NERON, V. SCHMITT

"Experimental and computational investigation of the vortex flow over a swept wing."
Eds. B. LASCHKA, R. STAUFENBIEL, Proceedings 14th ICAS Congress. AIAA, New York.

[4] M. BORREL, J.L. MONTAGNE

"Numerical Study of a Non-Centered Scheme with Application to Aerodynamics."
7th AIAA Comp. Fluid Dynamics Conference Cincinnati July 15-17th 1985.

[5] B. VAN LEER

"Flux-Vector Splitting for the Euler Equations."
Lecture notes in Physics. Vol.170, 1982, PP.507-512.

[6] L. CAMBIER, W. GHAZZI, J.P. VEUILLOT, H. VIVIAND

"A Multidomain Approach for the Computation of Viscous Transonic Flows by Unsteady Type Methods."
In Recent Advances in Numerical Methods in Fluids, Vol.III, W.G. Habashi, Ed. Viscous Computational Methods, Peneridge Press, 1984.

BREAKING THE I/O BOTTLENECK
WITH THE CRAY SOLID-STATE STORAGE DEVICE

Kent P. Misegades
Cray Research, Inc.
Mendota Heights, Minnesota   USA

SUMMARY

The use of the CRAY Solid-state Storage Device as a secondary memory
device for large-scale problems in computational fluid dynamics is reported.
The high data transfer rate and low access time of the SSD essentially
eliminates the I/O-wait penalty.   Hardware details, user impact (minimal,
the SSD is used like a conventional disk drive), several examples of its
use, and the relative performance of I/O types are discussed.

INTRODUCTION

For many problems in the field of computational fluid dynamics, the size of
computer main memory limits mesh resolution, thus limiting configuration
complexity and/or solution accuracy. The usual remedy to this limitation
is the use of secondary memory, i.e. disk drives. Figure 1 gives a simple
example of this concept.

```
          PROGRAM SSD
          COMMON /BIG/ ARRAY(300,200,100)
C            ARRAY HAS 6 000 000 ELEMENTS.
C            THE 20 DATASETS ON SECONDARY MEMORY ARE
C            USED TO GIVE 120 000 000 ELEMENTS.
          DO 40 IOFILE = 20,39
            DO 30 I = 1,300
              DO 20 J = 1,200
                DO 10 K = 1,100
                  ARRAY(I,J,K) = RANF()
10              CONTINUE
20            CONTINUE
30          CONTINUE
            WRITE (IOFILE) ARRAY
            REWIND IOFILE
40        CONTINUE
          DO 80 IOFILE = 20,39
            READ (IOFILE) ARRAY
            REWIND IOFILE
            DO 70 I = 1,300
              DO 60 J = 1,200
                DO 50 K = 1,100
                  ARRAY(I,J,K) = ARRAY(I,J,K) ** 2.
50              CONTINUE
60            CONTINUE
70          CONTINUE
            WRITE (IOFILE) ARRAY
80        CONTINUE
          STOP
          END
```

Fig. 1 Example used to demonstrate the use of secondary memory.

The temporary files used are assigned to CRAY DD-49 disk drives, capable
of reading and writing data at 10 Mbyte/sec. Execution of this simple
example on a CRAY X-MP/48 supercomputer (4 processors with 8 Mword total
64-bit main memory) resulted in a ratio I/O WAIT TIME / CPU TIME of 11:1.
Clearly, it is the transfer rate that causes a bottleneck between the
mainframe and the secondary storage. Ratios of 5-20 are common in real
application codes.

THE SOLID-STATE STORAGE DEVICE

First introduced in 1982, the CRAY Solid-state Storage Device is a very
fast random-access device for secondary storage. It is available with
either 32, 64, or 128 million 64-bit words of volatile, high-density MOS
memory. Access time is less than 50 microseconds, and data can be
transferred at a rate of 100 Mbyte/sec (X-MP/1), 1000 Mbyte/sec (X-MP/1
and 2), or 2000 Mbyte/sec (X-MP/4). The SSD can also be used on CRAY-1M
or CRAY-1S supercomputers. Figure 2 shows schematically the size, shape,
and position of an SSD relative to the 1, 2, and 4 processor X-MP mainframes.
The SSD cabinet consists of four columns arranged in 90° arc, requiring
24 square feet (2.3 square meters) of floor space. The power supplies are
hidden by the benchlike extensions arranged around the base. The liquid
refrigerant cooling system is similar to that of the X-MP mainframe.

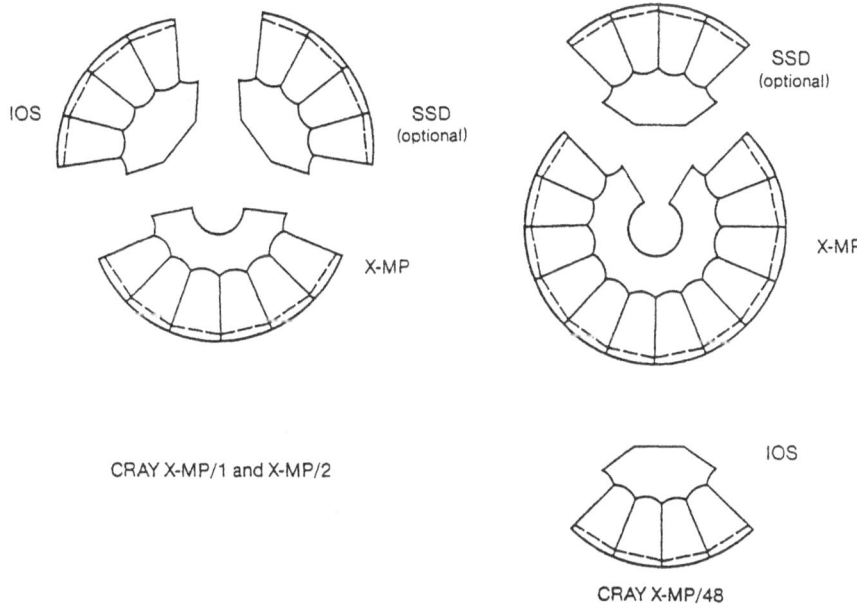

Fig. 2   The CRAY Solid-state Storage Device relative to the CRAY
         X-MP mainframe.

USER IMPACT OF THE SSD

For the user, the SSD is treated like a disk. Datasets are logically identical to those on disk storage, providing all standard FORTRAN I/O statements. No source program changes are required to access the SSD; just the addition of one parameter in the ASSIGN statements of the COS job control stream is needed. Figure 3 shows the COS statements used when accessing a disk and the SSD.

for DD-49 disk assignment

ASSIGN,DN=ANYNAME,A=FT10,DV=49-A1-26.

for SSD assignment

ASSIGN,DN=ANYNAME,A=FT20,DV=SSD-0-20

Fig. 3   COS statements used when accessing a disk or the SSD.

The device name appearing in the [DV=] parameter is a local, installation-dependent name. If only disk drives are desired, the parameter DV can be omitted. The operating system will then automatically assign datasets to the available disks. Should a user's dataset requirement at one point exceed his available SSD space, data overflows automatically onto conventional disks. In a multiple-CPU environment, simultaneous access to the SSD by more than one job is possible.

EXAMPLES

In the simple example of figure 1, using a 128 million word SSD resulted in a ratio I/O WAIT TIME / CPU TIME of 0:1, as compared to 11:1 using CRAY DD-49 disks. Since its introduction in 1982, various SSDs have shown significant benefits in such areas as weather forecasting, seismic processing, reservoir modeling, finite element analysis, as well as computational fluid dynamics. Below are two examples from such CFD applications.

SHANG 3D CODE

Joe Shang and Steven Scherr of the USAF Wright Aeronautical Labs recently reported on the solution of the flow field around a complete aircraft. Using a version of the SHANG 3D code, they computed the viscous flow around the Martin Marietta X-24C-10D lifting body. A mesh of 475,200 nodes was used. The computation was performed on the NASA Ames CRAY X-MP/22, of which 1 million 64-bit words of main memory and 8 million words of SSD memory were available to the user. In this investigation, 120 streamwise stations (pages) were divided into twenty data blocks of six pages each [1]. Each block of data was assigned to a data set of 213,840 words. Two blocks, or twelve pages, of data could be simultaneously resident in main memory at one time. Individual data blocks were rotated into main memory in sequence as required by the two-page lagging cycle of the MacCormack explicit method used. Standard FORTRAN synchronous, unformatted, blocked READ and WRITE statements were used. The observed wall clock time increase due to this I/O procedure was 23% of the total CPU time. This compares quite favorably to a degradation of approximately 500% experienced on the CYBER-205.

DORNIER/JAMESON 3D EULER CODE

During the summer of 1984, Jef Dawson of Cray Research and Stefan Leicher
of Dornier GmbH demonstrated the value of the SSD in a very large scale
CFD simulation. The DORNIER/JAMESON 3D Euler code is a finite volume
method using a Runge-Kutta 5-2 algorithm (5 R-K stages, 2 filtering calls
per iteration), enthalpy damping, local time-stepping, and implicit
smoothing. The case investigated was the steady, transonic, high angle of
attack flow about a delta wing having a biconvex airfoil (the so-called
DILLNER delta wing). The mesh, having 2.5 million nodes was divided into
16 blocks, one of which just filled the 8 million word main memory of a
CRAY X-MP/48 (4 processors, 8 million words of 64-bit main memory). At
every iteration, each of the 16 blocks, which just filled the 128 million
word SSD, was read into main memory using the unformatted, asynchronous
I/O statements BUFFER IN and BUFFER OUT. Each block was actually transferred
from and to the SSD twice in one iteration: once in the time step subroutine,
and once in the flow solver. 500 iterations of the code required 8.3
hours of wall clock time for 7.9 hours of CPU time on one processor of the
X-MP/48.

I/O TYPES ON CRAY SUPERCOMPUTERS

Whether a CRAY disk or a CRAY SSD is used for data transfer, the type of
I/O statements used is critical to achieving low wall clock times. The
possible types of I/O used on CRAYs (in order of decreasing speed):

I.   Asynchronous, unformatted, unblocked
II.   Asynchronous, unformatted, blocked
III. Synchronous, unformatted, unblocked
IV.   Synchronous, unformatted, blocked
VI.   Synchronous, formatted, blocked

How to call for each of these I/O types:

SYNCHRONOUS   Normal I/O statements

ASYNCHRONOUS   To read in the first 600 elements of array B:
                BUFFER IN(U,0) (B(600))

FORMATTED   Normal FORTRAN format specifiers

UNFORMATTED   WRITE (6) B

BLOCKED   Normal I/O statements

UNBLOCKED   In JCL:
                  ASSIGN,DN=ANYNAME,A=FT10,U.
                In FORTRAN source:
                  WRITE (10) B
                  where B is length 512 words

The relative performance of different I/O types is shown in the example of
writing out an array having 100000 octal elements. See Table 1.

TABLE 1. Relative Performance of Different I/O Types

TYPE	RELATIVE TIME
ASYNCHRONOUS, UNFORMATTED, UNBLOCKED	1
ASYNCHRONOUS, UNFORMATTED, BLOCKED	4.6
SYNCHRONOUS, UNFORMATTED, UNBLOCKED	1
SYNCHRONOUS, UNFORMATTED, BLOCKED, USING "WRITE A"	39
SYNCHRONOUS, UNFORMATTED, BLOCKED, USING IMPLIED DO	212
SYNCHRONOUS, UNFORMATTED, BLOCKED, USING WRITE IN A DO LOOP	28600
SYNCHRONOUS, FORMATTED, BLOCKED, USING "WRITE A"	37100
SYNCHRONOUS, FORMATTED, BLOCKED, USING IMPLIED DO	40000
SYNCHRONOUS, FORMATTED, BLOCKED, USING WRITE IN A DO LOOP	57500

As Table 1 shows, synchronous, formatted I/O is 37000-57000 times slower than asynchronous, unformatted, unblocked I/O. Note that, in many codes much computation can be performed while an asynchronous READ or WRITE is being carried out. The relative times shown in Table 1 do not reflect this advantage of asynchronous I/O. Further information regarding I/O types can be found in reference [2].

CONCLUSION

The CRAY Solid-state Storage Device essentially eliminates the I/O wait penalty incurred by using secondary memory in large-scale computations. For the user, switching from a disk to the SSD requires merely the addition of the DV parameter in the COS ASSIGN statement. It has also been shown that the choice of I/O type is extremely important, with synchronous, formatted, blocked I/O being the slowest type, and asynchronous, unformatted, unblocked the fastest.

REFERENCES

[1] "Navier-Stokes Solution of the Flowfield Around a Complete Aircraft", J.S. Shang and S.J. Scherr, December, 1984.

[2] CRAY FORTRAN (CFT) REFERENCE MANUAL, SR-0009.

SIMULATING 3D EULER FLOWS ON A
CYBER 205 VECTOR COMPUTER
by
P.J. Koppenol
National Aerospace Laboratory
A. Fokkerweg 2, Amsterdam
THE NETHERLANDS

SUMMARY

A computational method for solving the 3D Euler equations is studied. The
method is based upon an upwind flux-difference splitting scheme by Osher,
exhibiting an implicit mechanism for numerical viscosity, in connection
with an explicit time-marching finite-volume technique. The computer program is developed to run efficiently on both a scalar computer and the
Cyber 205 vector computer. Demands made by the necessity of vectorizability
of the code, on algorithm, data-structuring, and the code itself, are discussed. Also, the large data sets involved in 3D calculations, appear to
impose severe claims on central-memory size, I/O devices and line connections.
The method is tested for a transonic and supersonic quasi-two-dimensional
channel flow. The Euler model is found to give an accurate simulation of
aerodynamic phenomena in the channel.

1. INTRODUCTION

During the last decade, computing power has risen to a level on which flow
models based on the Euler equations can be used to compute flows around
three-dimensional geometries. Especially the introduction of the vector
computer (VC), together with the available mathematical techniques that
were developed in the same period, allowed the simulation of Euler flows,
on a potentially industrially interesting scale, with acceptable turn-
around times. Numerical techniques for solving hyperbolic systems of equations in general, and Euler equations in specific, have been a topic of
research since Godunov launched his famous method in 1959. Especially during the last five years, a large effort has been put into the development
of techniques for solving the Euler equations, resulting in a continuous
stream of papers. A promising class of schemes are the upwind techniques,
which split the physical information flow in some way, depending on the
direction of the information flow. Two famous subclasses of upwind techniques are those based on flux-difference splitting (e.g. Osher) and the
ones based on flux-vector splitting (e.g. Van Leer). With the introduction
of the first vector computer, a new generation of special purpose computers
became available: computers specially designed to perform large scale computations, and allowing high calculation speeds. In order to use these computers efficiently, it is necessary to tune the structure of the algorithm,
the data organisation, as well as the final coding to the vector arithmetic
of the available computer. However, we should carefully balance out the
performance of the code, e.g. concerning computation speed, and the readability, the modifiability and the transportability of it. This requires
well-trained users; "vector computing requires vector thinking".

2. THE EULER EQUATIONS: A HYPERBOLIC SYSTEM OF CONSERVATION LAWS

The Euler equations can be written in a differential form as

$$\partial_t U + \partial_u E . \partial_x U + \partial_u F . \partial_y U + \partial_u G . \partial_z U = 0, \tag{1}$$

in which

$$U = \begin{pmatrix} \rho \\ \rho u \\ \rho v \\ \rho w \\ e \end{pmatrix}, \quad E = \begin{pmatrix} \rho u \\ \rho u^2 + p \\ \rho uv \\ \rho uw \\ u(e+p) \end{pmatrix}, \quad F = \begin{pmatrix} \rho v \\ \rho vu \\ \rho v^2 + p \\ \rho vw \\ v(e+p) \end{pmatrix}, \quad G = \begin{pmatrix} \rho w \\ \rho wu \\ \rho wv \\ \rho w^2 + p \\ w(e+p) \end{pmatrix}. \tag{2}$$

U is the state vector with components: mass, momentum in x, y and z-direction, and energy. E, F and G are the flux vectors in x, y and z-direction respectively. x, y, and z are Cartesian coordinates. p is the static pressure:

$$p = (\gamma - 1) . (e - \frac{\rho}{2} (u^2 + v^2 + w^2)),$$

$\partial_u E = \partial E / \partial U$ etc. are Jacobian matrices of E with respect to U. This set of equations is known to be hyperbolic.
In one space dimension, the eigenvalues of the Jacobian matrix are the direction tangents in the (x,t) plane for the propagation in time of infinitesimally small disturbances. The curves along which these disturbances are transported are called characteristics.
The building blocks for solutions of hyperbolic systems of conservation laws are shocks, contact discontinuities and simple waves.
A shock is a discontinuity surface, stable to small disturbances, across which there is material transport. The stability is obtained by characteristics that converge into the shock. A contact discontinuity is a discontinuity surface across which there is no material transport. Simple waves are continuous solutions of (1) corresponding to exactly one eigenvalue. Rarefaction and compression waves belong to this class of solutions.
In order to clarify some steps that will be taken later on, we will make a short trip through phase space. The phase space is the subset of R^m that contains all realizable values of the state vector U.
Simple waves and contact discontinuities that separate two constant states U_1 and U_r in physical space, correspond to paths connecting two points in phase space. When we travel along this path from U_1 to U_r in phase space, we pass all intermediate states that we also pass while travelling from U_1 to U_r in physical space. When the simple wave corresponds to a certain eigenvalue λ_i of the Jacobian matrix considered, it can be shown that the path in phase space is everywhere locally tangential to the eigenvector $\underline{r}_i$ belonging to λ_i.
The contact discontinuity surface is a discontinuity with the characteristics on both sides of it, in time-space coordinates, parallel. This means that, along a path in phase space, connecting the two states at both sides, the eigenvalue belonging to the wave, say λ_k, does not change, implying

$$\nabla_u \lambda_k \cdot \underline{r}_k = 0 . \tag{3}$$

Functions of U with the property that their gradient in phase space is everywhere normal to a specific eigenvector are called Riemann functions. Thus, along a path chosen such that a given eigenvector is tangential to it, the Riemann functions related to this eigenvector are invariant.

A basic flow problem to be solved is the evolution during one time step of a single simple discontinuity in one space dimension, the Riemann problem:

$$U(x;t=t_0) = \begin{array}{l} U_1 \text{ if } x < 0 \\ U_r \text{ if } x \geq 0 \end{array} \qquad (4)$$

It is proven in [1] that, provided two states U_1 and U_r are sufficiently close, the solution of their Riemann problem consists of m+1 constant states, $U_0 \ldots U_m$, in physical space, separated by m primitive solutions (shocks, simple waves and contact discontinuities), while $U_0 = U_1$ and $U_m = U_r$.

3. DISCRETIZING PHYSICAL SPACE, THE STATE VARIABLES AND THE EQUATIONS

We assume that there exists a smooth one-to-one mapping of a rectangular box in computational space with Cartesian coordinates (ξ,η,ζ) onto the flow domain in physical space with Cartesian coordinates (x,y,z). This rectangular box is partitioned into cubical cells, parallel to the axes which map into hexahedral cells in the flow domain in an orderly fashion. The number of cells in the computational grid in the directions ξ, η, ζ are I, J, K respectively. The mapping should be such that problem boundaries are mapped onto computational block boundaries.

Since physical space is discretized by hexahedral cells, only the coordinates of the cell vertices in physical space need to be stored; these are $x_{ijk} = x(\xi_i;\eta_j;\zeta_k)$, $y_{ijk} = y(\xi_i;\eta_j;\zeta_k)$ and $z_{ijk} = z(\xi_i;\eta_j;\zeta_k)$ with $i \in \{0;\ldots; I\}$, $j \in \{0;\ldots; J\}$ and $k \in \{0;\ldots; K\}$. From these coordinates the areas of the lateral faces, the cell volumes and the vectors normal to the faces can be determined. We approximate, in each cell and over each time step, the state by the volumetric average value over the cell.

A discretization for the one-dimensional scalar conservation equation for cell i,

$$\partial_t \int_{\text{cell } i} U(x;t)\ dx + \int_{\substack{\text{boundary} \\ \text{of cell } i}} E(U).dS = 0, \qquad (5)$$

is given the form

$$\frac{U_i^{n+1} - U_i^n}{\Delta t} + \frac{E_R - E_L}{\Delta x} = 0 . \qquad (6)$$

E_L and E_R are numerical fluxes at the boundaries for which we have $E_L = E_L$ $(U_{i-1};U_i)$ and $E_R = E_R (U_i;U_{i+1})$ while, for consistency, $E_L (U_i;U_i)$ and $E_R (U_i;U_i)$ are equal to $E(U_i)$; the physical flux at U_i. Consider a one-dimensional control volume. Define at every cell boundary fluxes E^+ and E^-, corresponding to waves moving in positive or negative direction respectively, such that e.g. $E_L = E_L^+ + E_L^-$. The reason for this will become clear soon. Rewrite (6), by applying flux-difference splitting, to

$$\Delta x . \frac{U_i^{n+1} - U_i^n}{\Delta t} = -\{E_R - E_L\} , \qquad (7)$$

$$= -\{(E_R^+ - E_L^+) + (E_R^- - E_L^-)\}, \qquad (8)$$

following Osher, we write this as

$$= - \left\{ \int_{U_{i-1}}^{U_i} \partial_u E^+ . du + \int_{U_i}^{U_{i+1}} \partial_u E^- . du \right\}. \tag{9}$$

Here $\partial_u E^{\pm}$ is the Jacobian of the flux $E^{\pm}$ with respect to U. From the assumption of hyberbolicity, it follows that $\partial_u E$ may be factored into the product of a diagonal matrix D, and the matrix of eigenvectors T, thus

$$\partial_u E = T \; D \; T^{-1} \quad . \tag{10}$$

The diagonal elements of D are known to be the eigenvalues of $\partial_u E$. Now, $\partial_u E^+$ is defined as

$$\partial_u E^+ = T \; D^+ \; T^{-1} \quad , \tag{11}$$

where the diagonal elements of D^+ are $D_{ii}^+ = \max(\lambda_i ; 0)$, $\tag{12}$
$\partial_u E^-$ may be defined likewise.
The right-hand side of expression (9) may be rewritten to

$$\Delta x \; . \; \frac{U_i^{n+1} - U_i^n}{\Delta t} = - \tfrac{1}{2} \{ E(U_{i+1}) - E(U_i) \} + \tfrac{1}{2} \{ \int_{U_i}^{U_{i+1}} |\partial_u E| . dU - \int_{U_{i-1}}^{U_i} |\partial_u E| . dU \} \tag{13}$$

in which $|\partial_u E| = \partial_u E^+ - \partial_u E^-$. The second term of the right-hand side behaves as an elliptic operator, which automatically introduces the appropriate numerical viscosity.
Formula (13) has an advantage compared to formula (9): we need only determine the first of the two integrals and of the fluxes, since the other ones follow from the previous point.

4. THE APPROXIMATE RIEMANN SOLVER

By approximating the initial distribution by constant states in the cells, we introduce a Riemann problem at every lateral face. These Riemann problems will be solved by means of Osher's approximate Riemann solver. The first integral in (9) is determined by integrating $\partial_u E^+$ along a path between the end points U_{i-1} and U_i, in phase space. The path of integration Γ consists of a sequence of subpaths Γ_i. Each subpath Γ_i corresponds to an eigenvalue λ_i of $\partial_u E$, and is chosen in the subspace spanned by the eigenvector(s) corresponding to this eigenvalue. With the implicit function theorem, it may be shown that, given the order of the eigenvalues along Γ, the path is unique, provided that the endpoints of Γ are close enough in some sense. With this choice of path, the integral may be evaluated as follows:

$$\int_{U_{i-1}}^{U_i} \partial_u E^+ . dU = \sum_{k=1}^{3} \int_{\Gamma_i} \partial_u E^+ . dU \tag{14}$$

$$= \sum_{k=1}^{3} \int_{\Gamma_{i,\lambda_k} \geq 0} dE \ . \tag{15}$$

The integral has been reduced to simple flux differences over the parts of the subpaths, on which the corresponding eigenvalues are positive.
The generalization of the flux-difference evaluation of (9) in general coordinates is straightforward, when we realize that the fluxes used up to now, were fluxes normal to the cell faces considered. For an arbitrarily orientated cell face in three space dimensions, with unit normal vector $n = (n_x \ n_y \ n_z)^T$, the normal flux $\hat{E}$ is given by $n.\underline{F} = n_x.E + n_y.F + n_z.G$. With (2) this becomes

$$\hat{E} = \begin{pmatrix} \rho\hat{u} \\ \rho u\hat{u} + p.n_x \\ \rho v\hat{u} + p.n_y \\ \rho w\hat{u} + p.n_z \\ (e+p)\hat{u} \end{pmatrix} \quad ;$$

here, $\hat{u}$ is the velocity component normal to the cell face. Now, $\partial_u \hat{E}$ and $\partial_u \hat{E}^+$ can be defined analogously to $\partial_u E$ and $\partial_u E^{\pm}$, and the complete analysis given above is valid for $\hat{E}$ also. The final discrete conservation equation in general coordinates reads

$$U_{ijk}^{n+1} = U_{ijk}^{n} - \frac{\Delta t}{\Delta \xi} \cdot \left\{ \int_{U_{i-1,jk}^{n}}^{U_{ijk}^{n}} \partial_u \hat{E}^+.dU - \int_{U_{ijk}^{n}}^{U_{i+1,jk}^{n}} \partial_u \hat{E}^+.dU + \hat{E}(U_{i+1,jk}^{n}) - \hat{E}(U_{ijk}^{n}) \right\}$$

$$- \frac{\Delta t}{\Delta \eta} \cdot \left\{ \int_{U_{i,j-1,k}^{n}}^{U_{ijk}^{n}} \partial_u \hat{F}^+.dU - \int_{U_{ijk}^{n}}^{U_{i,j+1,k}^{n}} \partial_u \hat{F}^+.dU + \hat{F}(U_{i+1,jk}^{n}) - \hat{F}(U_{ijk}^{n}) \right\}$$

$$- \frac{\Delta t}{\Delta \zeta} \cdot \left\{ \int_{U_{ij,k-1}^{n}}^{U_{ijk}^{n}} \partial_u \hat{G}^+.dU - \int_{U_{ijk}^{n}}^{U_{ij,k+1}^{n}} \partial_u \hat{G}^+.dU + \hat{G}(U_{ij,k+1}^{n}) - \hat{G}(U_{ijk}^{n}) \right\}, \tag{16}$$

where $\hat{E}$, $\hat{F}$, $\hat{G}$ are the fluxes in the local one-dimensional problems at the cell faces in ξ, η and ζ directions, and computational space has been discretized with cubes of size $\Delta\xi.\Delta\eta.\Delta\zeta$.
Formula (16) and (15) are used in the implementation of the Riemann solver. They constitute a system of equations that is conceptually the same for the three coordinate directions and can be efficiently solved on a vector computer, as will be shown in section 8.

5. BOUNDARY TREATMENT

At the boundaries of a computational grid, it is in general not possible to adopt exactly the same difference formulas as those in the inner field. The special conditions that must be met here, may be incorporated by adapting the difference scheme or by continuing the grid in a special manner across the boundary. The physical conditions encountered in our problem are the zero-transpiration flux condition, (partially) prescribed in/outflow, and periodicity. These conditions must be modelled in our difference expressions in such a way that no artificial reflection of waves or other mechanisms that affect convergence and solution are introduced. In as much as we adopt a finite volume approach, the most natural way to account for boundary conditions is to place cell boundaries of the computational space onto the physical-space boundaries. For flexibility, vectorizability (next section) and simplicity reasons, we introduce virtual cells, just outside the physical boundaries. After each time step, the state vectors in these cells are updated such that, during the next time step, the required boundary condition will be fulfilled. With this procedure, we also achieve that the flux differences over cell faces on the flow boundaries are computed in the same way as those at cell faces in the inner flow. The implementation of the periodicity condition at the channel walls is trivial, and the zero-transpiration flux condition at the lower and at the upper boundary of the channel is realized by enforcing reflection at these walls.
We tested two approaches for treating the in/outflow boundaries.
In the first approach, all five state variables in the virtual boundary cells are kept fixed, some equal to true boundary values, the others equal to guessed values, hopefully corresponding to the desired steady solution values. This is called overspecification. The philosophy behind it is that the Osher scheme will automatically choose the appropriate subpaths for the integral. However, fixing state variables to erroneous values will affect the form of the path of integration between the inner and the virtual boundary cell and thus the resulting flux-difference expression. We expect that it is necessary for the guessed boundary values to be close to the steady-state values. The second approach uses the theory of characteristics to determine the boundary conditions on a given time level from prescribed fixed (boundary) conditions and from the states in the inner boundary cells on the previous level. This characteristic boundary procedure fits smoothly into Osher's methodology concerning Riemann-problem approximations. The approach corresponds to the first approximation of absorbing boundary conditions by Engquist and Majda [9]. They show that this treatment is maximally dissipative, weakly reflective and gives a well-posed problem.

6. VECTOR ASPECTS IN PROGRAMMING

For the calculations on fine grids, we used a Cyber 205 vector computer, situated at SARA computer centre in Amsterdam. The number of floating point operations to be performed for reaching steady state is $O(10^{11})$. This requires about one day on an average scalar computer, while the Cyber 205 can do it in less than one and a half hour.
The reason why a vector computer (VC) can be so very fast, compared to a scalar machine, is that it can perform several operations simultaneously in one clock cycle. The clock cycle is the smallest unit of time for a computer. For the 205, this clock cycle is 20 ns (20×10^{-9} seconds).
Every clock cycle, different operations are performed simultaneously by

so-called functional units. These units are capable of performing specific operations, there are e.g. special units for addition, for multiplication and for logical operations, that can work independently.

The cycle time determines the maximum speed of a VC. Since the Cyber 205 can perform two (different) floating point operations per cycle, this results in a maximum speed of 100 MFLOPS.

It is clear from figure 6.1, that the units in the pipe are only operating efficiently, if the data flow through the pipe is not interrupted. That is, if the same operations have to be performed on long vectors of data, stored in a set of contiguous memory locations. When a standard Fortran do-loop can be (re-)written such that it is a series of simple operations on "vectors", this loop is said to be "vectorizable".

For a loop to be vectorizable, some constructions should be avoided. These are e.g. relational operators, loop-dependent subscripts, "calls", recursion, vector lengths larger than 65535, and branching if-statements. By performing operations in vector mode, a considerable speed increase can be obtained compared to scalar mode. This is illustrated with figure 6.2. In order to obtain between 25 and 50 percent of the maximum speed of a VC, more than 90 percent of the dynamic code must be vectorizable. (The number of dynamic-code lines is defined as the sum of the number of code-lines times the number that this line is executed during program execution. This is the counterpart of the number of static-code lines, where every line is counted once).

Vectorizing a program, which means that as many do-loops as possible are written in such a way that they can be performed in vector mode, can be done in three ways. The easiest way is by using a compiler option, which activates the compiler to automatically rewrite the vectorizable do-loops into vector code. The compilers abilities to do this, however, are limited. Another possibility is to use a special preprocessor e.g. VAST (Vector and Array Syntax Translator), to do the vectorization. The third possibility is vectorize the program explicitly.

On the Cyber 205, several Fortran 77 extensions are available to vectorize a program explicitly; that is, by using a special syntax, the VC is forced to perform these operations in vector mode. A first option for this, is the explicit vector syntax: a shorthand for do-loops. A way to perform an operation on a whole vector in vector syntax, is to use vector functions. Most standard Fortran functions have their vector analogon, while other additional utility functions, such as the dot product, exist. A low level means of enforcing vectorization, is to use assembly language statements in the Fortran code; the Fortran version available on the Cyber 205, FTN 200, supports special "Q-calls" for operations, explicit I/O etc. They will not be illustrated here. An important feature on the 205, is "controlled store". This is a way to vectorize loops with if-statements which would normally not vectorize. A bitvector is set, with "ones" at places where the operation on another vector must be performed. Then the operation is performed on the whole vector, but the results are stored only at places where this bitvector is 1. Although the results on the places where the bitvector is zero are thrown away, there is in general a speed increase compared to the scalar mode, since now the operations can be performed vectorwise.

When a do-loop cannot be auto-vectorized just because the data are not contiguous in memory, there is a procedure that enables us to explicitly vectorize it as yet. The memory locations that have to be processed are gathered in a contiguous auxiliary array. Then the operations are performed, after which the data are scattered into memory again. Special instructions are available for performing "gather" and "scatter" operations on vectors.

Now that we have seen how a VC processes its data and why we should write vectorizable programs to fully exploit the VC, some remarks must be made about code qualities. These are the readability, the (trans)portability, the modifiability, and the correctness of our code. We assume that our code is at least correct, in the sense that it calculates what we want it to. But what about the other aspects? From what we saw in the above, it is clear that a very efficient code is not portable, not even transportable (it can be run on another machine, but with small modifications). When advanced tricks are used, we may even loose the readability of the program. Nowadays, practical codes are a compromise between efficiency and the other code qualities. It depends on the specific application and circumstances, whether we emphasize one or the other.

For our code, readability, modularity and, especially, transportability are important aspects. Therefore, we have written it in such a way that most of the code is scalar code that automatically vectorizes. Some portions are explicitly vectorized, using vector functions, explicit vector syntax and controlled store, but these parts were at the same time coded in a scalar version. Less than 10 percent of the static code lines are explicitly vectorized lines. In the compilation run, we can activate either the scalar or the vector-code parts in a machine-independent way. We did not use the special Q-system calls, in order to retain readability. The result is a short, readable code, with a large amount of modularity, that can be run efficiently on a scalar machine and on a vector machine without changes.

7. DATA HANDLING

Three-dimensional flow calculations on fine grids involve very large data sets. In this section, we discuss the resulting problems, how to reduce the data size, and how we organized our data.

How rapidly the data size increases with the grid dimensions is shown in table 7.1. For this table, we assumed that the solver requires five state variables, three coordinates and thirty-eight auxiliary real numbers, in every grid point.

TABLE 7.1
Total data size in kilo-words for several grids.

GRID	18x3x8	66x3x26	130x3x50	128x32x32
DATA SIZE IN KWORDS	20	237	897	6029

Since available CM size of the Cyber 205 is only 800 Kwords, it is obvious that, for fine grids, the data set cannot be kept in CM completely. As a consequence, I/O has to take place between central memory and background memory during the calculations. This can result in shocking I/O time/ CPU time ratios.

To reduce the I/O time, the data size has to be kept as small as possible. To achieve this, we adopt a blocking technique: computational space is partitioned into blocks, consisting of P planes of cells, parallel to the $\eta 0\zeta$ plane, see figure 7.1. P is at least 1 and most I, and should be chosen such that one block easily fits into CM. During each time step, the blocks are updated one by one, in central memory.

Since 28 of the 38 auxiliary arrays do not require I/O, the amount of data to be I/O-ed can be considerably reduced. Details about the data flow in the algorithm are presented in the next section.

The amount of data, to be available in CM for calculations, may also be reduced by reducing P, since the length of the arrays required in central memory for the calculations, is equal to the block length: PxJxK, see figure 7.2.

These, and other effects of varying P have different consequences for program efficiency. In section 11 we will see that there is an optimal value for P, with respect to CPU time.

The blocking technique is only a partial solution to an I/O problem. An additional action that can be taken to reduce I/O overhead, is to update a block in CM for several time steps, keeping the states in the adjacent blocks constant. Experiments have shown, [8], that this procedure may reduce I/O waiting time by a factor of up to 25, without affecting convergence rates. Large central memory or very fast I/O devices may also offer a solution to the I/O problem. Finally, solutions may be found in adapting the algorithm or arranging for asynchronous I/O (I/O is performed parallel to the calculations).

CPU time may also become a bottleneck in 3D calculations. To reduce the CPU time, we may also use software or hardware approaches: by paying more attention to vectorization (by adapting one subroutine, we saved 30 % of CPU time), by numerically adapting the time-consuming Osher solver, by implementing a multiple grid technique, by using fast assembly language calls (special Q-calls), or by using a multi-processor computer with a very small clock cycle. It should be stressed that it is wise to accept some loss of efficiency on behalf of readability and modifiability.

A third consequence of large data sizes occurs when a remote VC is used. Transport times, necessary to get in-house fine-grid results via a slow commercial data connection can be very large. Table 7.2 shows that, in the absence of a fast data channel, the only practical means for transporting time-evolution results on fine grids is by tape, which can be teasingly slow. Since a VC works most efficient on large sets of contiguous memory locations, we stored all arrays one-dimensionally. The cells are counted as depicted in figure 7.3. Virtual boundary cells are incorporated, both in the arrays and in the block-wise calculations. The length of the state-variable array is IxJxK, the length of the auxiliary arrays is PxJxK and the length of the arrays containing the coordinates of the cell vertices is (I+1)x(J+1)x(K+1).

A consequence of this way of numbering the cells is that, except for the boundary treatment, the vectors of data in all operations are contiguous, independent of any of the coordinate directions. So the operations may be vectorized without gathering and scattering the data. Only the offsets of the vectors are coordinate-direction dependent.

TABLE 7.2

Theoretical transport times for one time level of state variables

GRID LINE SPEED	128x32x32	196x48x48
1200 b.	9,7 hours	1,4 days
9600 b.	1,2 hours	4,2 hours
50 Kb.	14 min	47 min

8. THE ALGORITHM

The flow solver is part of a flow solver system of which the general structure is shown in figure 8.1. This system embraces several autonomous programs (the encircled items) and data sets (the parallelograms). The flow
solver is the link between the preprocessing part and the postprocessing
part.
The main algorithm is depicted in figure 8.2. It is structured in such a way
that extra features, such as Runge-Kutta time stepping, or a higher-order or
implicit method, may be implemented without major problems. By including the
boundary cells in the one-dimensional arrays, we automatically incorporate
the information that comes from these cells, and all operations vectorize.
But, we have to prevent that the flux-difference contributions are added to
residuals at virtual-boundary cells. This is done by using bitvectors, with
zeros at places corresponding to virtual-boundary cells, see figure 8.3.
The structure of the Riemann solver, as it is formulated in (16), is given
in figure 8.4. The controlled-store option is also used when the flux-dif-
ferences along the subpaths in phase space are determined: the flux vectors
are evaluated simultaneously at a given intermediate point, for all cells of
a block. By using bitvectors, we filter out the flux-vector terms in those
cells for which this vector value does not occur in the flux-difference expression.
In figure 8.5, the data flow during the execution is visualized. From this
figure, we may see how the blocks of data are processed by the CPU, and it
is therefore a means of analyzing possible I/O adaptions for the algorithm.
The final code containing the whole 3D Euler solver comprises only about 900
lines, comment lines not counted. These lines are divided over 18 subroutines. The Riemann solver takes about 45 percent of the code. We found that
a large number of routines indeed gives a flexible code that is easy to
modify. The code was designed, and debugged in six man-months. In the
future, the code will be used to test improvements to the Riemann solver.

9. AVAILABLE COMPUTER CONFIGURATION

Not only the kinds of computers that are used in a project, such as vector
and scalar computers, but also some aspects of the computer configuration
may show up to be important.
The computer configuration that was available, is sketched in figure 9.1.
The work station is situated at the NLR-site in Amsterdam. Three different
computers are used, which means: three different operating systems, three
different access procedures, two different text editors (the CYBER 205 is a
batch computer), and different library-maintenance tools.
Since the postprocessing has to be done on the CYBER 180/855 at NLR, large
result arrays must be transported from the CYBER 205 to the CYBER 855. Even
for our relatively small mesh size of 130x3x50 points, transfer times of
several hours per time level of state variables were observed. During this
study, the only alternative to the slow 1200 baud connection between the
CYBER 825 at SARA and the CYBER 855 at NLR, is to drive a tape to and fro:
this alternative was not used. An obvious solution would be to install a
fast data line.
An additional problem results from using a dial-up line for communication
between the work station and the front-end computer; it is the noise that
spoils edit sessions and frequently results in key-board locks.
From our experience, we conclude that the lack of fast and troublefree data

connections and user-friendly machine-machine interactions restrict and slow down numerical research for 3D fluid dynamical research on remote VC's.

10. PLOT FACILITIES

It is very difficult to get insight into the distribution of a variable depending on three space dimensions; we have to restrict ourselves to two-dimensional subsets of the 3D domain, or to specific values of the variable of interest.
We experienced that, for numerically investigating a 3D solution, the following plot facilities are desired: contour plots of a variable on grid-planes in computational and physical space, displaying a variable as a function of two coordinates of computational space (variation plots), and contour bodies in physical and computational space. These possibilities should be embellished by multi-colour, hidden-line techniques and shading.
We found that a combination of at least two different ways of plotting a variable is the fastest way to get insight into the 3D distribution of it. In our project, we use contour plots in combination with variation plots, which works well. For more complicated 3D flows, more advanced possibilities must be developed, however.

11. RESULTS

The results presented in this section include both the numerical experiences with the Osher solver on our test problem, and the informatics results on the CYBER 205.
The test problem we have chosen, is a transonic or supersonic flow through a channel with a 4.2 % thick circular arc at the bottom, see figure 11.1. The transonic case was a test case for the GAMM workshop in 1979 [7]. Inflow and outflow conditions are prescribed as a ratio of .6235 for static downstream pressure to total upstream pressure. In isentropic flow, this corresponds to Mach .85.
Since the channel flow is two-dimensional, we need only three cells in span-wise direction: one inner cell surrounded by two virtual cells. In figure 11.1, the inner cells of our coarsest grid are depicted. Cell sizes in x and z-direction can be halved three times. Figure 11.2 shows a detail of the finest grid (130x3x50 cells).
The test cases of which results will be shown, are:

Case	Mach	Grid	N.time st.	Remarks
1	.85	18x3x8	2000	Oversp. at in- outflow bnd's.
2	.85	18x3x8	2000	Oversp. at inflow, charact. at outflow
3	.85	130x3x50	5000	Oversp. at in- outflow
4	3.	130x3x50	2000	Oversp. at inflow, charact. at outflow

That shocks are captured sharply, provided they are approximately aligned to the grid lines, becomes evident from figure 11.3. Here, the Mach number in the cell centres of a plane y = constant is plotted perpendicular to this plane. The figure shows that the shock has two interior cells.
The oblique-shock spreading, which is evident from the iso-mach lines in figure 11.4, is an imperfection in the Osher scheme: the implicit numerical viscosity is not switched off properly for discontinuities not aligned to

81

grid lines. The convergence plots of cases 1 to 4 are given in figure 11.5, Measured CPU times and the resulting estimated MFLOP rates are given in table 11.2.

TABLE 11.2

CPU times and MFLOP rates, for case 1 and a version of case 3.

Case	Vector length	N. of steps	CPU s	MFLOPS
1	432	2000	60	19
3	1950	2000	1800	26

The MFLOP rates are determined by estimating a lower bound for the total number of floating point operations, and then dividing by the total number of CPU seconds. They also include subroutine calls and other overhead. The corresponding rate of data processing for case 3 is 4.1×10^{-5} seconds per grid point and per time step.

The coarse grid was first tested on the CYBER 180/855 at NLR. A speed increase with a factor of 15 was observed when we ran the code on the CYBER 205. The influence of decreasing the number of planes P in a block, is shown in table 11.3. Conflicting effects resulting from decreasing P (section 6), lead to the optimal value P=13.

TABLE 11.3

Influence of P (number of planes in a block) on vector length in the operations, on the size of the controllee file in 512-word blocks, and on the CPU seconds per time step.

Grid	P	vector length	contr. file in blocks	CPU seconds per step
130x3x50	130	19500	2554	1.02
	65	9750	1806	.987
	26	3900	1358	.915
	13	1950	1206	.897
	1	150	1070	1.27
			CM = 1612	

Autovectorization on the CYBER 205 can take place via the preprocessor VAST and by the FTN200 compiler. When using VAST, one must be careful to check whether the code produced by VAST is indeed faster than the original one. Generally, we use both VAST and the compilers optimization facilities. The VASTorized and compiler-optimized scalar code was found to be a factor of 3 slower than the VASTorized and compiler-optimized code with the explicit vector statements: the vector version of our code. So in this case, the explicit vectorization has paid off.

12. CONCLUSIONS

The upwind technique based upon the flux-difference splitting scheme of Osher introduces the appropriate numerical viscosity (section 3). His approximate Riemann solver allows a physical interpretation (section 4) and, though it involves many tests, can be easily vectorized (section 6). The approximate Riemann solver captures shocks that are aligned to grid lines, with at most two interior points, and without oscillations. But when the shock is not approximately aligned to the grid, it smears out considerably (section 11). Although the scheme is first-order accurate, the results obtained with it show much detail and are an invitation to further research in order to resolve current deficiencies. These concern: slow convergence (expensive), spreading of oblique shocks, large CPU time per grid point per time step for the solver, and the need to incorporate weakly or non-reflecting boundary conditions.

Further research is necessary to understand why the overspecification-boundary treatment works (it is very attractive from an implementation point of view), and how to eliminate entropy losses at cells where the grid-derivatives are discontinuous.

To make a code highly efficient on a specific machine, means that important code properties such as readability, modifiability and transportability, may be affected. However, it appeared to be possible to write an Euler solver in correct, readable, transportable and modifiable Fortran, that still is efficient on a vector computer, (section 6).

3D numerical simulations generally involve large data sets. To account for the I/O problems that may result from these data sizes, measures have to be taken; for example, code or algorithm adaptions. The blocking technique, described in section 7, is a reasonable solution but may not be sufficient.

3D simulations also require a huge number of floating point operations for reaching steady state. This makes the use of a VC together with a cheap and vectorizable solver inevitable (section 7). Performing 3D calculations on a remote VC without a high-speed data connection, slows down and restricts the development and use of 3D-flow simulation codes (section 7).

3D numerical experiments require new 3D plot facilities, and a well developed 3D-imaginative faculty (section 10).

REFERENCES

[1] Lax, P.D.: "Hyperbolic systems of conservation laws and the mathematical theory of shock waves", Reg. Conf. Series in Applied Math., SIAM 1973, no. 11.

[2] Harten, A., Lax, P.D., Leer, B. van: "On upstream differencing and Godunov-type schemes for hyperbolic conservation laws", SIAM review no. 25.1, (1983).

[3] Osher, S., Solomon, F.: "Upwind difference schemes for hyperbolic systems of conservation laws", Math. of Comp., Vol. 38 (1982) pp. 339-374.

[4] Osher, S.: "Numerical solution of singular perturbation problems and hyperbolic systems of conservation laws", Math. studies no. 47 (1981), North Holland Publ. Comp., pp. 179-204.

[5] Smoller, J.: "Shock waves and reaction diffusion equations", Grundlehren der Math. Wissenschaften, 258 Springer Verlag (1983).

[6] Koppenol, P.J.: "Hyperbolic systems of conservation laws - theoretical and numerical aspects", NLR Memorandum IW-84-009 U, (1984).

[7] Rizzi, A., Viviand, H.: "Numerical methods for the computation of inviscid transonic flows with shock waves", Notes on numerical Fluid Mechanics, Vol. 3, Vieweg (1979).

[8] Radespiel, R., Kroll, N.: "Progress in the development of an efficient finite volume code for the three-dimensional Euler equations", DFVLR, IB129-84/20, (1985).

[9] Engquist, B. Majda, A.: "Absorbing boundary conditions for the numerical simulation of waves", Math. of Comp., Vol. 31, No. 139 (1977), pp. 629-651.

[10] Koppenol, P.J.: "3D Euler flow simulations on a Cyber 205 vector computer", NLR TR 85054 U, (1985).

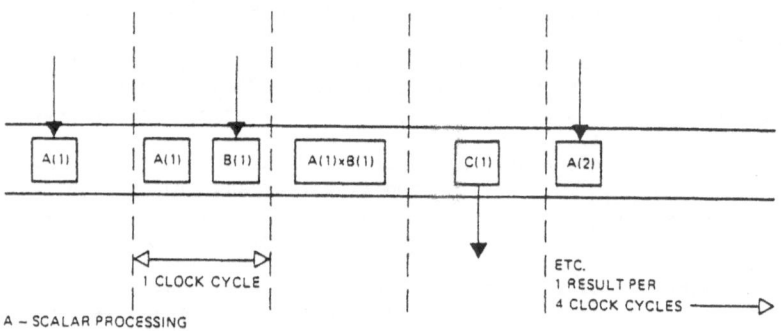

1 CLOCK CYCLE

ETC.
1 RESULT PER
4 CLOCK CYCLES ⟶

A – SCALAR PROCESSING

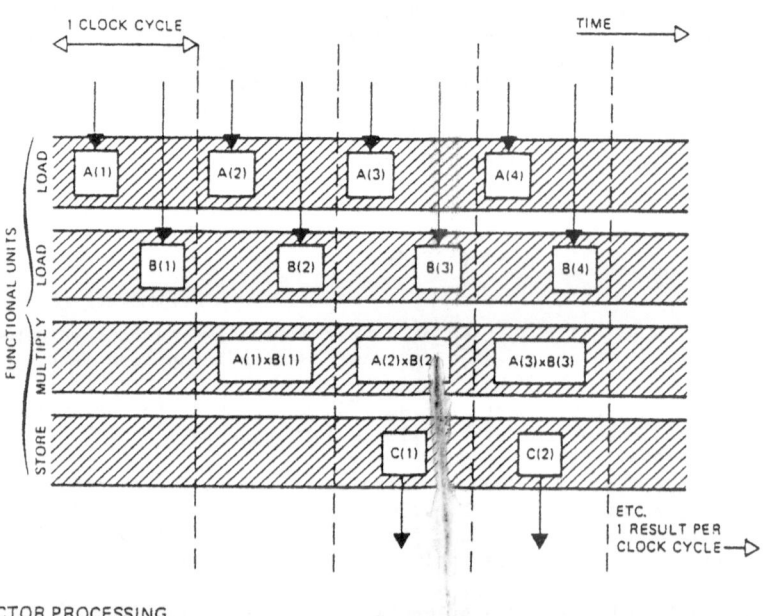

B – VECTOR PROCESSING

Fig. 6.1 Scalar versus Vector processing. The vector-processor part
consists of several functional units, able to work in parallel.
The figure is related to the loop

```
        DO 10  I = 1,500
          C(I)  = A(I)*B(I)
     10    CONTINUE
```

OPERATION	SCALAR MODE	VECTOR MODE
I = 1(1) N : SIN (A(I))	230 C	13 C
I = 1(1) N : ATAN 2 (A(I), B(I))	686 C	19 C

Fig. 6.2 Number of clock cycles (c) per operand for different operations
and modes (the start-up time has been neglected)

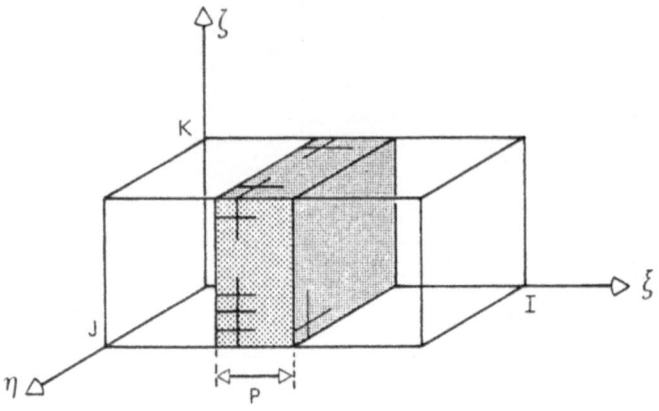

Fig. 7.1 Computational grid, divided into blocks that contain P planes of JxK cells

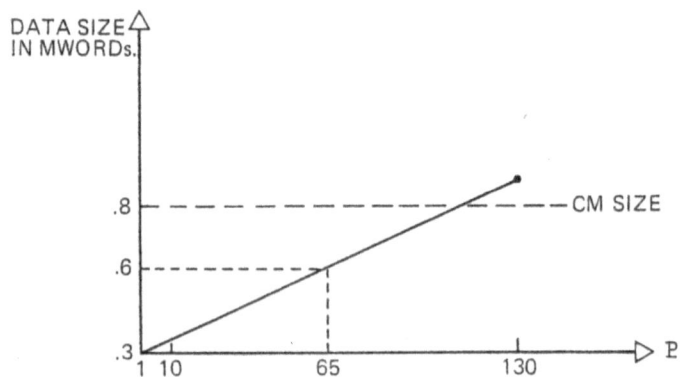

Fig. 7.2 Datasize for the 130x3x50 grid as a function of P, the number of planes in one block

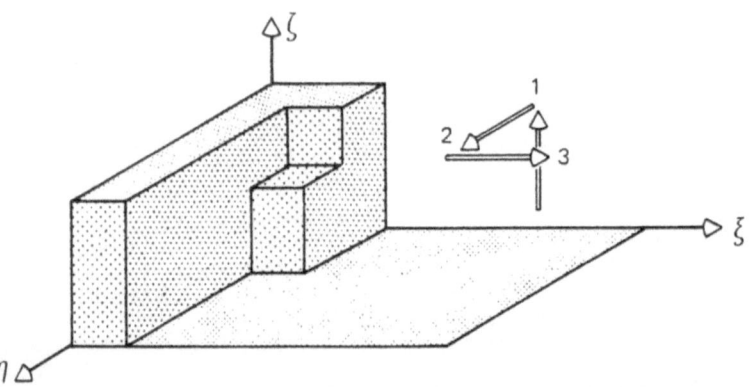

Fig. 7.3 Cell ordering. Starting at the origin, cells are counted, first in ζ direction (columns), then in η direction (planes) and finally in ξ directions

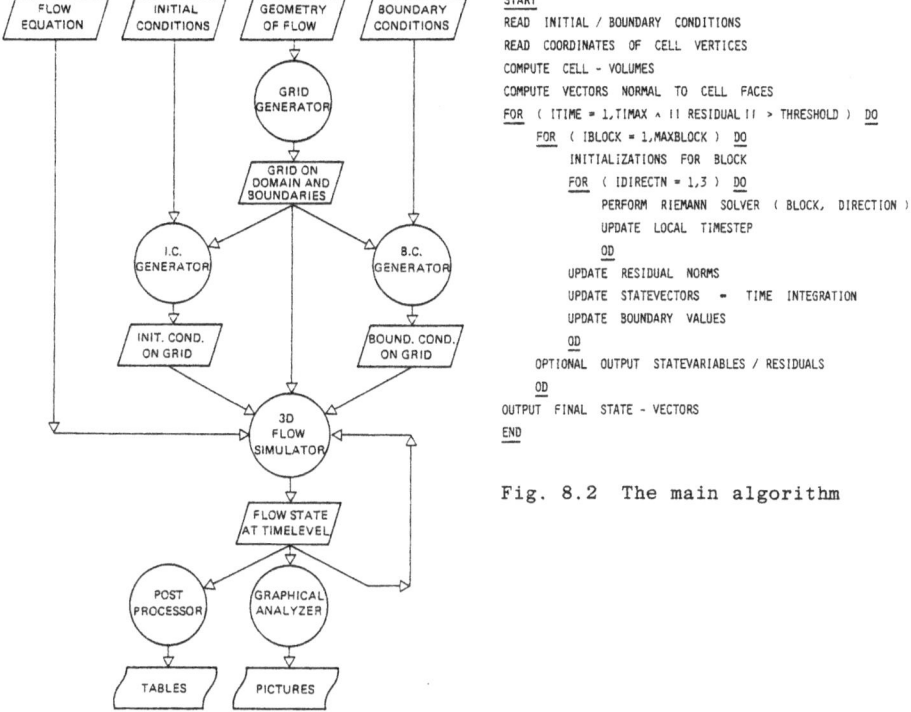

```
START
READ  INITIAL / BOUNDARY  CONDITIONS
READ  COORDINATES  OF  CELL  VERTICES
COMPUTE  CELL - VOLUMES
COMPUTE  VECTORS  NORMAL  TO  CELL  FACES
FOR  ( ITIME = 1,TIMAX ∧ || RESIDUAL || > THRESHOLD )  DO
     FOR  ( IBLOCK = 1,MAXBLOCK )  DO
          INITIALIZATIONS  FOR  BLOCK
          FOR  ( IDIRECTN = 1,3 )  DO
               PERFORM  RIEMANN  SOLVER  ( BLOCK,  DIRECTION )
               UPDATE  LOCAL  TIMESTEP
               OD
          UPDATE  RESIDUAL  NORMS
          UPDATE  STATEVECTORS  -  TIME  INTEGRATION
          UPDATE  BOUNDARY  VALUES
          OD
     OPTIONAL  OUTPUT  STATEVARIABLES / RESIDUALS
     OD
OUTPUT  FINAL  STATE - VECTORS
END
```

Fig. 8.2   The main algorithm

Fig. 8.1   Structure of flow solver system

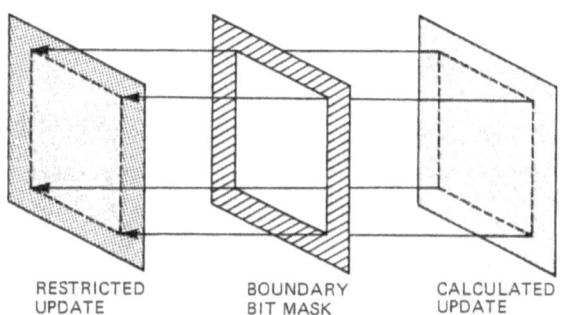

RESTRICTED   BOUNDARY   CALCULATED
UPDATE   BIT MASK   UPDATE

Fig. 8.3   Preventing the spoiling of states at virtual boundary cells,
           by using a bit mask

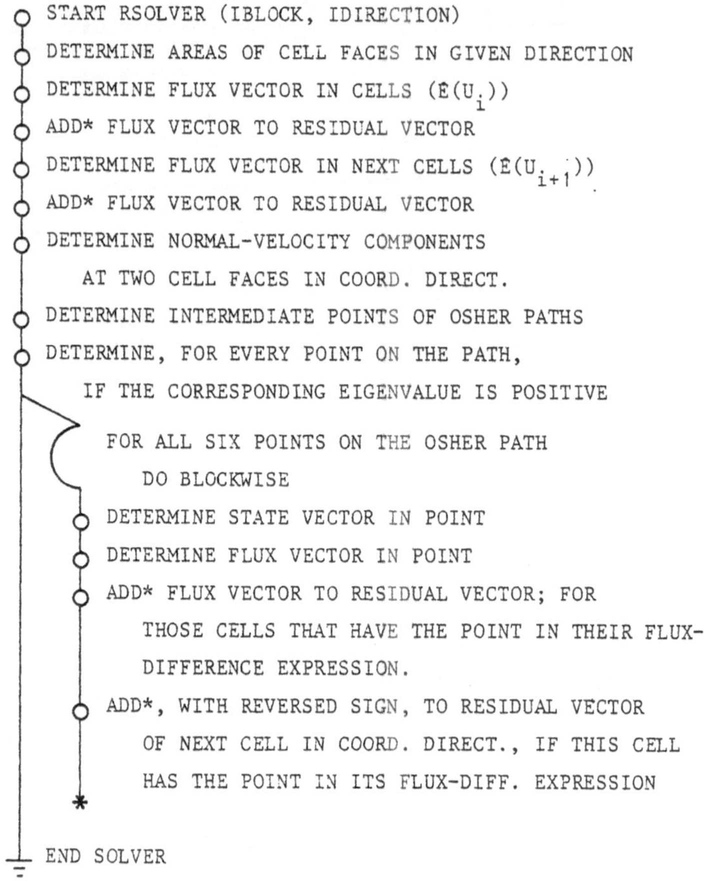

START RSOLVER (IBLOCK, IDIRECTION)

DETERMINE AREAS OF CELL FACES IN GIVEN DIRECTION

DETERMINE FLUX VECTOR IN CELLS $(\hat{E}(U_i))$

ADD* FLUX VECTOR TO RESIDUAL VECTOR

DETERMINE FLUX VECTOR IN NEXT CELLS $(\hat{E}(U_{i+1}))$

ADD* FLUX VECTOR TO RESIDUAL VECTOR

DETERMINE NORMAL-VELOCITY COMPONENTS
   AT TWO CELL FACES IN COORD. DIRECT.

DETERMINE INTERMEDIATE POINTS OF OSHER PATHS

DETERMINE, FOR EVERY POINT ON THE PATH,
   IF THE CORRESPONDING EIGENVALUE IS POSITIVE

   FOR ALL SIX POINTS ON THE OSHER PATH
      DO BLOCKWISE

DETERMINE STATE VECTOR IN POINT

DETERMINE FLUX VECTOR IN POINT

ADD* FLUX VECTOR TO RESIDUAL VECTOR; FOR
   THOSE CELLS THAT HAVE THE POINT IN THEIR FLUX-
   DIFFERENCE EXPRESSION.

ADD*, WITH REVERSED SIGN, TO RESIDUAL VECTOR
   OF NEXT CELL IN COORD. DIRECT., IF THIS CELL
   HAS THE POINT IN ITS FLUX-DIFF. EXPRESSION

END SOLVER

ADD*: ADD ONLY TO NON-BOUNDARY POINTS

Fig. 8.4 Dimensional flowchart of the Riemann solver.
          It closely follows formula (16)

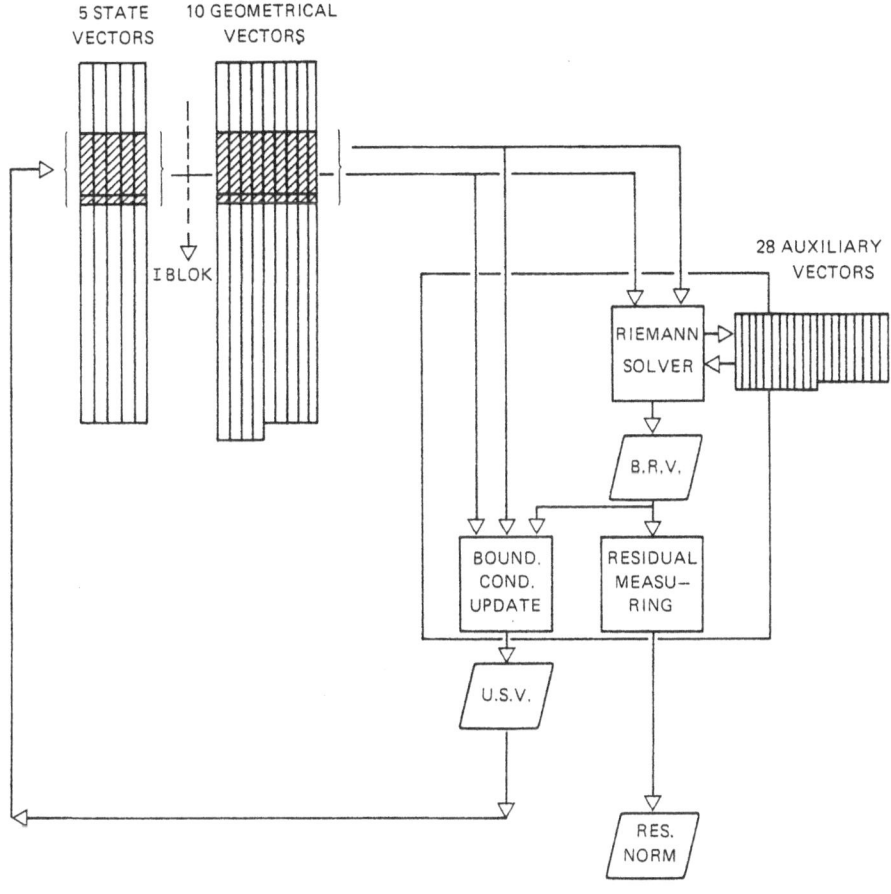

5 STATE VECTORS

10 GEOMETRICAL VECTORS

IBLOK

28 AUXILIARY VECTORS

RIEMANN SOLVER

B.R.V.

BOUND. COND. UPDATE

RESIDUAL MEASU— RING

U.S.V.

RES. NORM

B.R.V. - BLOCK - RESIDUAL VECTOR
U.S.V. - UPDATED STATE - VARIABLES

Fig. 8.5 Data flow diagram

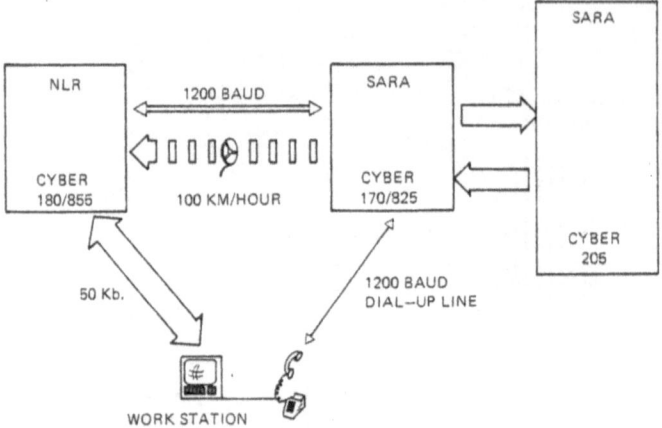

Fig. 9.1   Available computer configuration

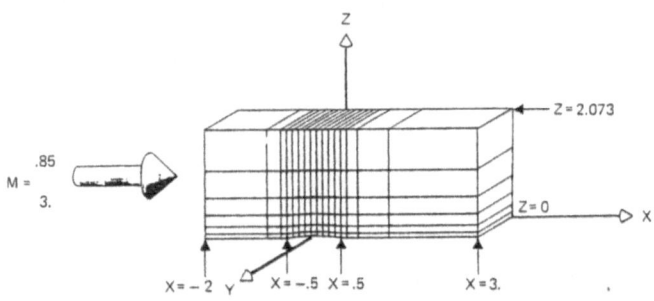

Fig. 11.1   Transonic or supersonic flow through a channel with a 4.2 %
thick bump at the bottom, between x = -.5 and x = .5. Inner
cells of the 18x3x8 grid are shown.
x = -2: inflow boundary, x = 3: outflow boundary,
z = 0 and z = 2.073: reflecting boundaries,
y = 0 and y = 1: periodicity boundaries.

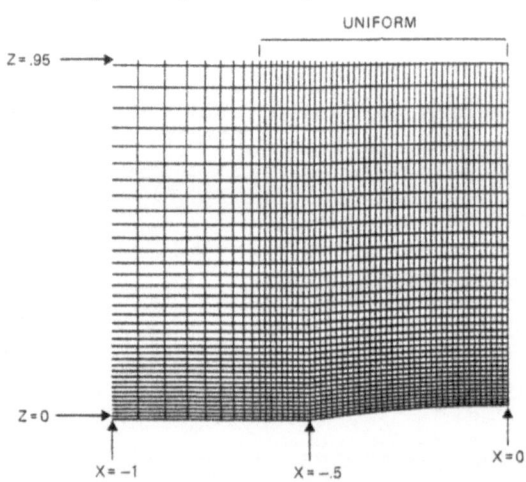

Fig. 11.2   Detail of 130x3x50 grid

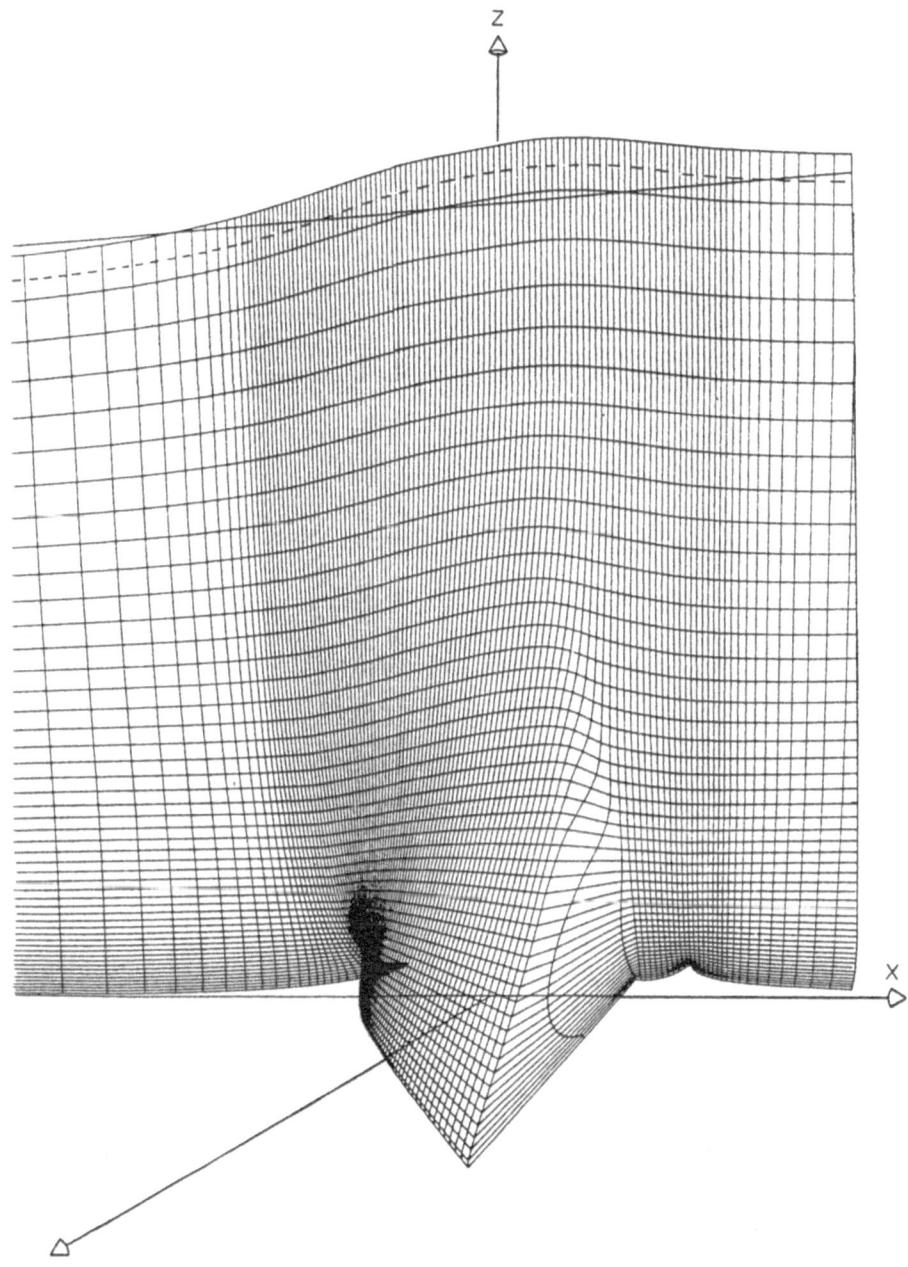

MACH NUMBER

Fig. 11.3 2D Mach number distribution
          -1.45 < x < 1
          .587 < M < 1.315

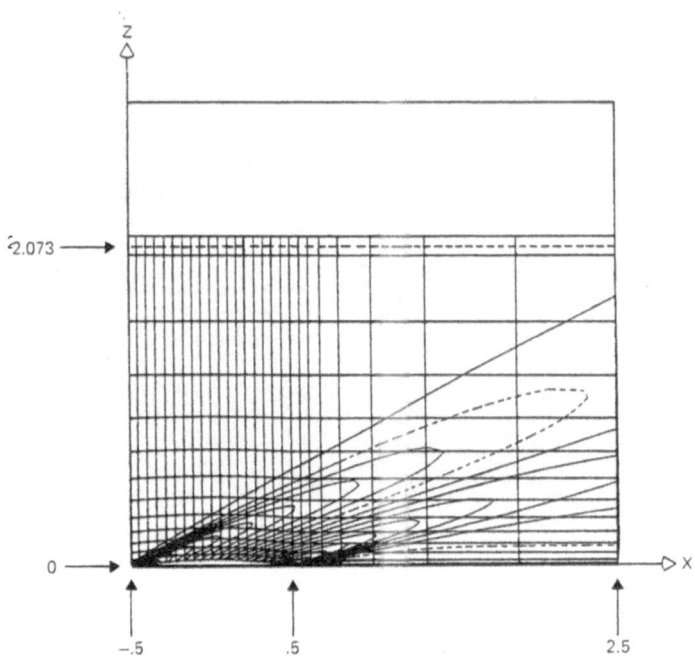

Fig. 11.4   2D-Iso-Mach lines. Smearing of shocks

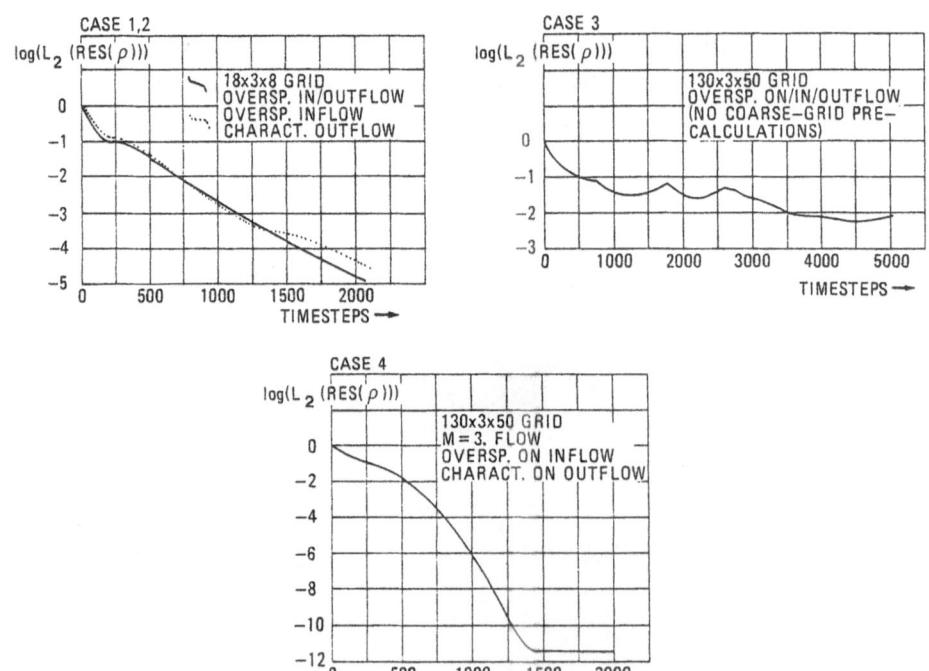

Fig. 11.5   Convergence histories for cases 1 to 4

VECTOR ALGORITHM FOR LARGE-MEMORY CYBER 205
SIMULATIONS OF EULER FLOWS

Arthur Rizzi[*]

FFA, The Aeronautical Research Institute of Sweden,
S-161 11 BROMMA, Sweden
and
J.P. Therre
Ecole Polytechnique Fédérale, Lausanne, Switzerland

SUMMARY

The paper reviews a finite-volume method for the large-scale numerical simulation of fluid flow and discusses the vector coding and execution of the procedure on the CYBER 205. With the proper structure given to the data by the grid transformation each coordinate direction can be differenced throughout the entire grid in one vector operation. Boundary conditions must be interleaved which tends to inhibit the concurrency of the overall scheme, but a stragey of no data motion together with only inner-loop vectorization is judged to be the best compromise. The computed example of transonic vortex flow separating from the sharp leading edge of a delta wing demonstrates the processing performance of the procedure. Vectors over 40,000 elements long are obtained, and a rate of over 125 megaflops, sustained over the entire computation, is achieved when the entire data set is resident in real memory. Attemps to use secondary memory, either explicitly or with virtual management, greatly degrades the performance.

INTRODUCTION

Realistic vortex flowfield simulations require three dimensional discrete models of significant size. The standard supercomputer typically has 1 or 2 million words of central memory. The user then has a choice, he either tailors the size of his mesh so that his entire data set fits into main memory, or he constructs a larger mesh and resorts to secondary memory to process the larger data set. Two drawbacks with the latter approach are that the data are not word addressable, and therefore must be buffered in and out, and that the transfer rates for secondary memory are very much slower. The usual strategy tries to overlap the arithmetic processing of in-core data with the I/O buffering of other data in secondary memory. Its success is strongly dependent on the particular algorithm and computer hardware, e.g. the buffer size, transfer rate etc. Our attempts with this approach on the CYBER 205 all met

---

[*] also adjunct prefessor of computational fluid dynamics, Royal Institute of Technology, Stockholm, Sweden

with heavy time penalties due to the slower secondary memory. It appears that our algorithm is much better balanced with the very high transfer rate to main memory on the CYBER 205, than it is with secondary memory. But we are not restricted to small mesh dimensions because fortunately the central memory of every CYBER 205 can be increased by a simple field upgrade to at least 8 M words. The recent construction of a 16-million-word memory for the CYBER 205 (or the NASA-VPS32) for example has allowed tests of one-million-grid-point models on a practical basis within reasonable elapsed times. This demonstrates that both the hardware as well as the software is up to the task of large-scale flowfield simulation. Therefore good virtual-memory-management techniques within a sufficient working set of real memory seems to be the only effective way at present to carry out such large-scale simulations. The other crucial requisite is a large band-width communication channel between the memory and the arithmetic unit. We describe how our algorithm makes maximum use of these two important features of the CYBER 205.

Accompanying the advent of such a powerful vector computer, however, is an increased discipline on programming technique and the recognition of data structure. To bring out what this actually means in practice we summarize the main features of the finite-volume method for flowfield simulation, and describe how the mesh topology directly determines the data structure. Then follows the vector coding of the finite-volume procedure to solve the Euler equations in three dimensions. The essential prerequisite for processing operands in the vector unit of the CYBER 205 is that they lie in contiguous locations in memory, and the single most crucial item in the vectorizing process is the design of the data-flow algorithm. The problem then is not how to use secondary memory, but rather how to vector process most effectively the data in real memory. If we disregard the boundaries, it becomes obvious after a little reflection that all of the data lie in contiguous locations in memory. When boundaries are included further thought must be given to the layout of the data, and where data are not contiguous, decisions must be made whether the extra effort to transpose the data to be contiguous (so-called data motion) is worth the increased speed gained by vector processing. Our particular choices in conjunction with the data-structure design lead to the following key features of the procedure: 1) Separate storage arrays are assigned for the dependent variables q, flux component F, and flux differences FD, 2) one extra unit is dimensioned for each computational direction to hold the boundary conditions, and 3) flux differences are taken throughout the entire field by off-setting the starting location of the flux vector F. In this way all of the work in updating interior points is exclusively vector operations without any data motion, accomplished at the expense of just a single scalar loop for the boundary conditions on one computational direction and a doubly-nested loop for another, while the boundary conditions for the third direction vectorize completely. In this way one can vectorize completely the calculation of the metric terms and eliminate all triply-nested scalar DO loops in taking the three dimensional differences - all with a minimum demand on central memory and no transposition of data in storage. Instead of relying on automatic vectorization by the compiler the algorithm is coded directly in vector syntax using explicit semicolon notation in vector arithmetic operations and bit vector constructs in place of scalar logical branching. The actual performance of the code, its megaflop rate together with the overall efficiency gained by vector processing is demonstrated by a computed example of transonic vortex flow separating from the leading edge of a delta wing at high angle of attack. The O-O type grid that discretizes the space surrounding the wing contains 192×56×96 cells, just over one million. The vector length of the variables, which spans a three-dimensional subset of the data set that results from this grid discretization, is over 40,000 elements long in this case. The entire data is processed through a scalar loop over this vectorized three-dimensional subset, a standard technique usually known as strip mining or domain decomposition.

FINITE VOLUME METHOD

The first step in carrying out a numerical solution is to discretize the flowfield by creating a mesh that spans the entire region. Eriksson's method of transfinite interpolation[1] constructs a boundary-conforming O-O grid around a 70 deg. swept delta wing, the so-called Dillner wing. An equivalent form of the Euler equations can be expressed as an integral balance of the conservation laws

$$\frac{\partial}{\partial t} \iiint q \; dvol + \iint \underset{\sim}{H}(q) \cdot \underset{\sim}{n} \; dS = \quad \begin{array}{c} \text{artificial} \\ \text{viscosity model,} \end{array} \quad (1)$$

where q is the vector with elements $[\rho, \rho u, \rho v, \rho w]$ for density and Cartesian components of momentum with reference to the fixed coordinate system x,y,z. The flux quantity $\underset{\sim}{H}(q) \cdot \underset{\sim}{n} = [q\underset{\sim}{V}+(0,\underset{\sim}{e}_x,\underset{\sim}{e}_y,\underset{\sim}{e}_z)p] \cdot \underset{\sim}{n}$ represents the net flux of q transported across, plus the pressure p acting on, the surface S surrounding the volume of fluid. The mesh segments the flowfield into very many small six-sided cells, in each of which the integral balance (1) must hold. The finite volume method then discretizes (1) by assuming that q is a cell-averaged quantity located in the center of the cell, and the discretized flux term $[\underset{\sim}{H} \cdot \underset{\sim}{S}]_{ijk}= [\underset{\sim}{H}(\mu_I q_{ijk}) \cdot \underset{\sim}{S}_I + \underset{\sim}{H}(\mu_J q_{ijk}) \cdot \underset{\sim}{S}_J + \underset{\sim}{H}(\mu_K q_{ijk}) \cdot \underset{\sim}{S}_K]$, where μ is the averaging operator, is defined only at the cell faces by averaging the values on each side, see Fig. 1. With these definitions and calling the cell surfaces in the three coordinate directions of the mesh $\underset{\sim}{S}_I$, $\underset{\sim}{S}_J$, and $\underset{\sim}{S}_k$, we obtain the finite-volume form for cell ijk

$$\frac{\partial}{\partial t} q_{ijk} + [\delta_I(\underset{\sim}{H} \cdot \underset{\sim}{S}_I) + \delta_J(\underset{\sim}{H} \cdot \underset{\sim}{S}_J) + \delta_k(\underset{\sim}{H} \cdot \underset{\sim}{S}_k)]_{ijk} \cdot \underset{\sim}{n} = \quad \begin{array}{c} \text{aritificial} \\ \text{viscosity model,} \end{array} \quad (2)$$

where $\delta_I(\underset{\sim}{H} \cdot \underset{\sim}{S}_I)= (\underset{\sim}{H} \cdot \underset{\sim}{S}_I)_{i+1/2}-(\underset{\sim}{H} \cdot \underset{\sim}{S}_I)_{i-1/2}=\delta_I f_I$ is the centered difference operator. To this the boundary conditions for the particular application must be specified. They occur at the six bounding faces of the computational domain and when included, we can write Eq.(2) as

$$\frac{\partial}{\partial t} q_{ijk} + FD = \begin{array}{c} \text{artificial viscosity model} \\ + \\ \text{boundary conditions} \end{array} , \tag{3}$$

where FD represents all of the flux differences. A more detailed description of the method is given in Ref. 2 which is the origin of the present computer program and Ref. 3 which describes its most recent development. With the spatial indices suppressed we integrate this last equation with the two-level three-stage scheme in parameter $\theta = 1/2$ or 1

$$\begin{aligned}
q^{(0)} &= q \\
q^{(1)} &= q^n + \Delta t\ FD(q^{(0)}) \\
q^{(2)} &= q^n + \Delta t\left[(1-\theta)FD(q^{(0)}) + \theta\ FD(q^{(1)})\right] \\
q^{(3)} &= q^n + \Delta t\left[(1-\theta)FD(q^{(0)}) + \theta\ FD(q^{(2)})\right] \\
q^{n+1} &= q^{(3)},
\end{aligned} \tag{4}$$

that steps the solution forward in time until a steady state is reached. The number of cells in a suitable grid around a wing can range from 50 thousand to several millions and the computational burden to solve Eqs.(3) and (4) repeatedly for say 1000 time steps is large and requires a supercomputer with the greatest speed and largest memory.

Algorithm (4) is explicit, i.e. the state q^n for each point known at time t is updated pointwise to the new state q^{n+1} by simple algebraic equations involving no simultaneous solution of equations or recursive relations. Therefore the updating of different points in the field are independent of each other and can be carried out concurrently. And it is this concurrency that has led to the reexamination of explicit algorithms for large computational problems with the aim to maximize the obtainable vector processing rate.

ALLOCATION OF COMPUTER RESOURCES

Before starting on the programming of the algorithm, we must lay down a strategy for using those computer resources that we have at our disposal. Figure 2 displays them as the instruction unit, the arithmetic unit, central memory, and peripheral storage. The size of the central memory sets an upper limit on the number of grid points in the discrete model. It may be possible to raise this limit somewhat by resorting to peripheral storage, but this too has its limits. The goal therefore is to strike a balance between work in the CPU and data transfer, which then leads to the central question of whether to save an intermediate result in memory or recompute it locally from the primary values. Because of its high communication rate between central memory and the CPU, the CYBER 205 favors recomputing as many intermediate variables as possible, and thereby maximize the total number of grid points. The allocation found to perform best is indicated in Fig. 3 where all of the 21 primary arrays are stored in real memory, and virtual memory was relied upon to manage the secondary arrays.

DATA STRUCTURE AND METHODOLOGY FOR VECTOR PROCESSING

To the CYBER 205 programmer a vector is just an ordered
list of data items whose elements (integer, real, complex or
even bits) must be stored in a contiguous set of memory locations. The number of elements on the list is called the vector length, unlike the mathematical definition. The simpliest
example of a vector is a one-dimensional FORTRAN array. At the
FORTRAN level a vector is designated by a descriptor that
contains the vector's starting address and length. The explicit descriptor takes the form

ARRAY (INDEX; INTEGER EXPRESSION)

and points to the vector whose first element is the element
residing in ARRAY at position INDEX and whose length is the
number given by INTEGER EXPRESSION. The nature of the vector
instruction in the CYBER 205 emphasizes rapid operations upon
contiguous cells in memory.

The goal is to process Eqs.(3) and (4) with a maximum
degree of vectorization but at the same time a minimum demand
on central memory, and in order to accomplish this we have to
devise a strategy for structuring the data. The finite-volume
method uses an ordered grid system that maps the physical
space to the computational space through an underlying mesh
transformation. It is this mapping that determines the overall
structure of the computational data, as shown in Fig. 4 for a
single-block O-O grid. Boundary conditions are set on the six
faces of the computational space (see Fig. 5). Consider the
flux array F(I,J,K) in the interior of the three-dimensional
computational domain shown in Fig. 6 to which the boundary
conditions will be added later to the beginning and end of
each coordinate direction. A three dimensional structure suitable for vector processing is correctly visualized as consisting of a collection of adjacent pencils of memory cells with
suitable boundary conditions. This leads to the natural grouping for the formation of differences in J by means of differencing adjacent pencils. The natural differencing in the K-
direction thus becomes the difference of a plane of adjacent
pencils. Differences in the I-direction are formed by use of
single elements offset one against the next within the pencil.
See Fig. 7 for these relationships which allow for the complete vectorization of all of the internal difference expressions. In the I-direction the contiguous group is the individual storage cell, in the J-direction it is the pencil of all
I cells for a given J and K, and in the K-direction it is the
plane of all I*J cells for a given K. Because of the ordering
of that data by the grid transformation, the total vector difference throughout the interior domain can then be formed by
subtracting the respective group from its forward neighbor
simply by off-setting the starting location of the flux vector
for each of the three directions. The inherent concurrency of
the algorithm now becomes apparent. In this way all of the
work in updating interior points is exclusively vector operations without any data motion, because of the ordering that

results from the structured grid. The problem remaining is how to interleave the boundary conditions into this long vector. They require different operations and interrupt the vector processing of the interior values. (See Fig. 5). One approach, in conjunction with the data-structure design, that overcomes this interruption of vector processing leads to the following key features of the procedure: 1) separate storage arrays are assigned for the dependent variables q, flux component F, and flux differences FD, 2) dimension one extra group in each direction for q(I+1,J+1,K+1) and F(I+1,J+1,K+1) in order to insert the forward boundary values in q and to provide inter-mediate scratch space during the flux computation which is then overwritten with the rear boundary condition on F. Four variables are updated at each boundary cell. These boundary conditions are set on every stage of each iteration. In each direction the starting boundary value is established first on the property q, and the ending value on the flux F. Now fol-lowing this procedure we can difference the entire three-dimensional flux field F together with its boundary conditions with one vector statement involving vectors whose length can be as large as the total number of nodal points in the grid. The variable q is then eventually updated as a function of the differences in F (i.e. FD), and so forth. (Fig. 8). Figure 8 is an attempt to illustrate this algorithm schematically for differences in the K-direction which can be computed with full vectorization and no data motion. To begin, the right boundary (i.e. last plane, K+1) values are loaded in q(1,1,K+1;M) in one vector statement where M= (I+1)* (J+1) is the number of elements in the plane. This value is picked up by the averag-ing operator μ when computing the leading or outward (i.e. K+1) flux for the Kth plane of cells. The left boundary (first plane) then must be set, and because there is scratch storage already there, we choose to enforce this directly on the flux F(1,1,1;M) entering the first plane of cells. Now all differ-ences in the K-direction are ready to be taken in one vector instruction simply by offsetting the starting location of the flux vector

   $FD(1,1,1;N) = F(1,1,2;N) - F(1,1,1;N)$.

Diagrammatically the data flow in the algorithm may be viewed as a series of steps starting with q and pyramiding up to the flux difference FD over the whole K-direction.

   A similar diagram could be drawn for the other two direc-tions J and I, but then instead of in whole planes the bound-ary values enter in pencils and individual isolated elements respectively of the vectors q and F, and the inherent concurr-ency of the algorithm degrades. The compromise we make is to avoid transposing these boundary values to achieve continguity and instead accept shorter vector length and even scalar loop-s. Thus a high degree of vectorization achieved with no data motion and a minimum requirement for storage - 21 three-dimen-sional arrays in all: 4 for geometry (x,y,z,vol), 8 for two levels of the dependent variables q, 3 altogether for tempor-ary values of pressure p, flux F and velocity flux QS and QT, 4 for the total differences FD of each of the four governing equations, and 1 for the local time step DTL. (Fig. 3).

98

The actual hardware for the CYBER 205 has a vector-length limit of 65 535 elements which is easily exceeded by the three-dimensional vector differencing scheme described above. The technique for circumventing this possible limitation is called strip-mining for the CYBER 205. Each portion of the data set is described as a three-dimensional subset (or slab) of the final data ensemble. This method creates internal boundary conditions between slabs which are accommodated within the computational procedure. A scalar DO-loop is used to process each slab in turn at an average vector length approaching 65 000 elements. This strip-mining technique is performed sequentially (slab by slab) in the CYBER 205 but with a machine having parallel possibilities it could be processed concurrently. The concept for strip mining is shown in Fig. 9.

The generalized mapping from data to slabs is managed via common block declarations within FORTRAN. The common blocks are grouped into large pages by the loader. The large pages are stored on the disk system across a series of independent temporary files, each of which is accessed and mapped into the executable space by the operating system. All of this is transparent to the user which means the code is transportable from one memory size to another without user intervention.

SIMULATED VORTEX FLOWFIELD

A prime motivation to solve the Euler equations is to simulate vorticity-dominated flows. One typical example is flow separating from the sharp edge of a low-aspect-ratio wing and rolling up into a primary vortex above the wing. The computation of one such aerodynamic flow of practical interest for the design of a combat aircraft in high-speed maneouver, supersonic flow $M_\infty=1.75$ $\alpha=18^\circ$ past a sharp-edged 70° swept delta wing, demonstrates how the code performs. The mesh used is an 0-0 type (Fig. 10) consisting of 193 nodes around the semispan section, 97 on both the upper and lower chord sections, and 57 from the wing outward to the far boundary, amounting to just over one million. Because of the high angle of attack vortex flow is expected, and Fig. 11 presents contour lines of computed vorticity magnitude $|\Omega|$ and Mach number in 3 selected spanwise grid surfaces (not planar) that clearly reveal the strong interaction with the leading edge, producing very high shear flow that then forms cross-flow shocks and vortex structures over the upper surface of the wing. Very complex wave interactions take place in the flow over the wing--coalescing waves growing into shocks together with expansion waves that interact with the vorticity produced by the shocks. The isoMach lines and total pressure contours in Fig. 12 on the upper surface of the wing show some of these phenomena. Another region worth studying is the wake just behind the trailing edge where a complex interactin between the vortices formed from the leading and trailing edges, the shear layers, and the shocks takes place as evidenced by the vorticity and isoMach contours just before the trailing edge and in the wake (Fig. 13 and Fig. 14).

The convergence history of this case during the course of the computation is given in Fig. 15 by plots of log of the residual error, lift and drag all versus the number of time cycles and CPU time on the CYBER 205. The computation begins with the freestream field on the fine mesh and after 1000 cycles and 8000 CPU seconds the residuals are reduced by practically 3 orders of magnitude for an average reduction rate 0.988. The performance of the computer for this case is summarized in Fig. 16. The code executed in 32-bit precision on a two-pipe machine and performed at a rate of 8 CPU microseconds/cycle/grid point. Since the number of operations per cycle per grid point is something like 1000, this means that the machine sustained a rate of more than 125 megaflops over the entire course of the computation. That is better than half the asymptotic rate of the machine and indicates the overall design of the vector algorithm. Because of the long vector length and the high degree of vectorization of the code, the processing speed on a 4-pipe CYBER 205 would be nearly double the ones quoted above.

FINAL REMARKS

The appearance of super-computers with super-size real memory is opening up new vistas for the fluids software engineer for the study of much more realistic and practical problems than he has been able to before. It also allows the computational physicist to study fundamental fluid phenomena that previously were inaccessible, for example the direct simulation of fluid instabilities and even of turbulent flow. The technology now exists to build the necessary hardware, and the computer manufacturers are encouraging us, the application specialists, to seize the opportunity and expand our discrete models to the full limits of today's hardware in order to discover the unknown. Up to now that seems to require real memory. Perhaps techniques will soon be demonstrated that show that secondary memory can also be effective. In any case we must show the boldness and the determination to take this challenging step. To take full advantage of the power of the new hardware requires a greater programming discipline than we have been accustomed to. It also means an active concern for the structure of the data and its flow through the algorithm. But the challenge will be met.

REFERENCES

[1] Eriksson, L.E.: Generation of Boundary-Conforming Grids around Wing-Body Configurations using Transfinite Interpolation, AIAA J., Vol. 20, Oct 1982, pp. 1313-1320.

[2] Rizzi, A.W. and Bailey, H.E.: Finite Volume Solution of the Euler Equations for Steady Three-Dimensional Transonic Flow. Proc. 5th Int'l Conf. Num Meth Fluid Dynamics, eds. A.I. van der Vooren and P.J. Zandbergen, Lecture Notes in Physics, 59, Springer Verlag, pp. 347-357, 1976.

[3] Rizzi, A.W. and Eriksson, L.E.: Computaltion of Flow Around Wings Based on the Euler Equations, Journal Fluid Mechanics, Vol. 148, Nov. 1984.

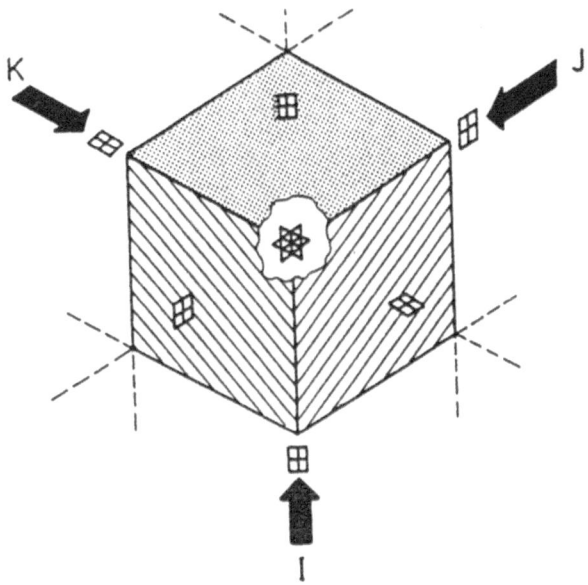

a) the cell averaged variable q is positioned
at the center of the cell.

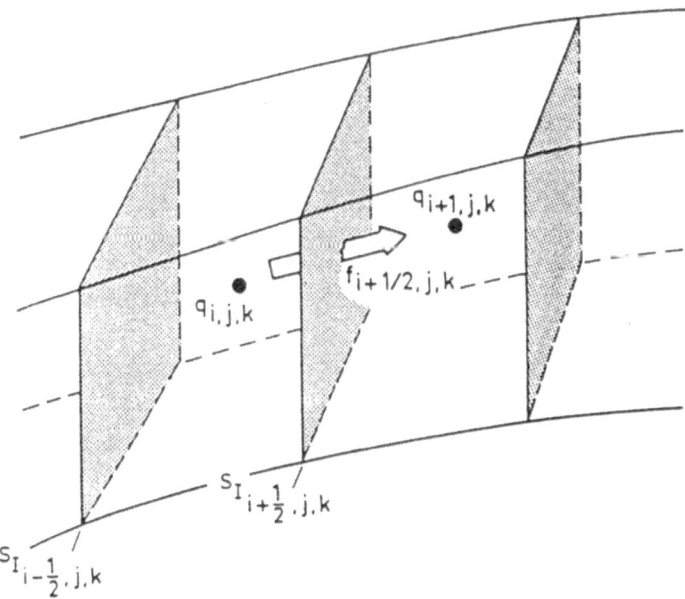

b) the flux F cross the faces of each cell is
the average of the fluxes obtained from
the two values of q that each face straddles.

Fig. 1   Hexahedrons discretize the domain with the finite
volume concept.

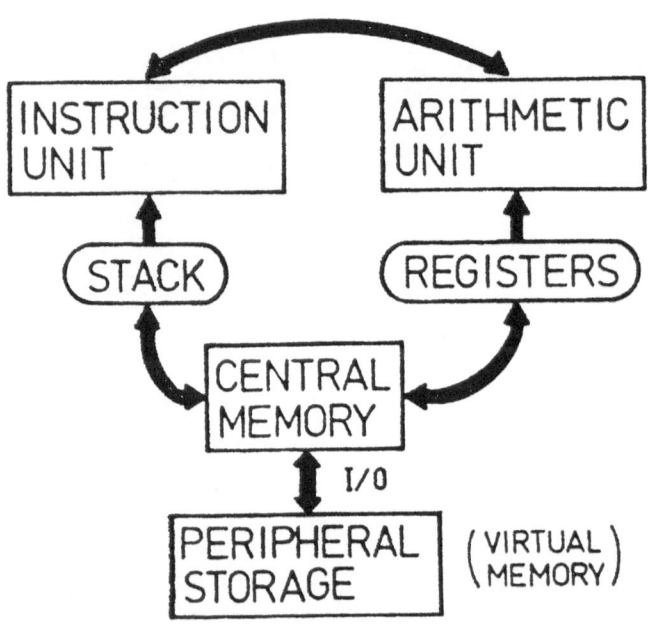

Fig. 2   The computer resources at our disposal must be allocated to yield a balance among the size of central memory, the power of the CPU, and the transfer times for instructions and data.

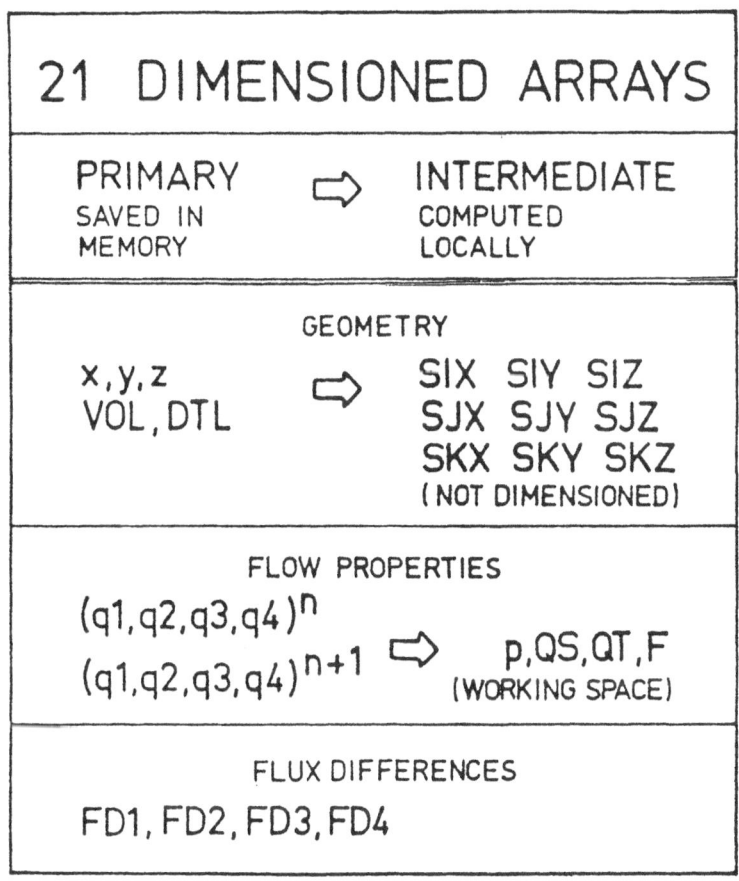

Fig. 3   Optimal balance of the CYBER 205 resources calls
          for storing 17 primary variables as large arrays
          together with 4 more working space making a
          total of 21 altogheter.

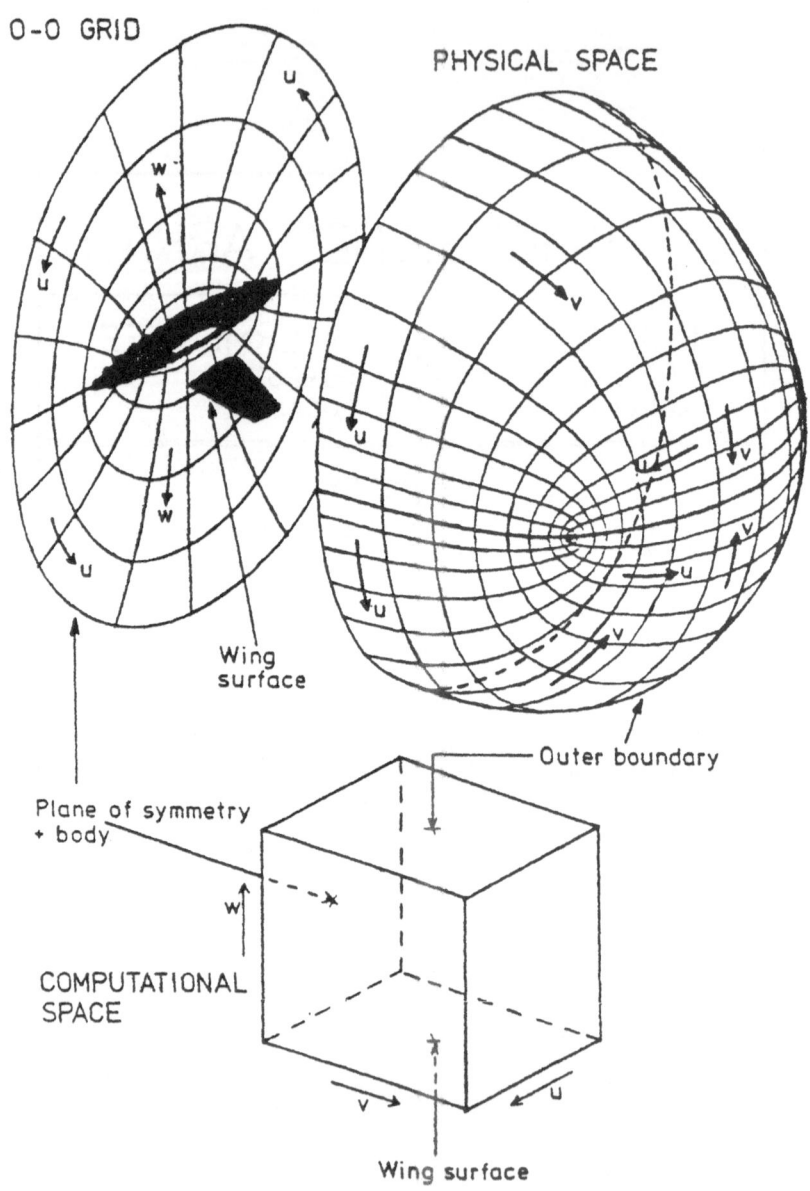

O-O GRID

PHYSICAL SPACE

Wing
surface

Plane of symmetry
+ body

Outer boundary

COMPUTATIONAL
SPACE

Wing surface

Fig. 4   Grid structure determines the ordering of FORTRAN
dimensioned arrays through the mapping of the
physical domain to the computational space. Any
vectorization technique can take advantage of this
structure.

Boundary Conditions

Three Types

- nonreflecting farfield

- periodic coordinate cuts

- solid surfaces
  1. zero flux-normal mom. eq.
  2. trailing edge Kutta condition?
     not needed if sharp

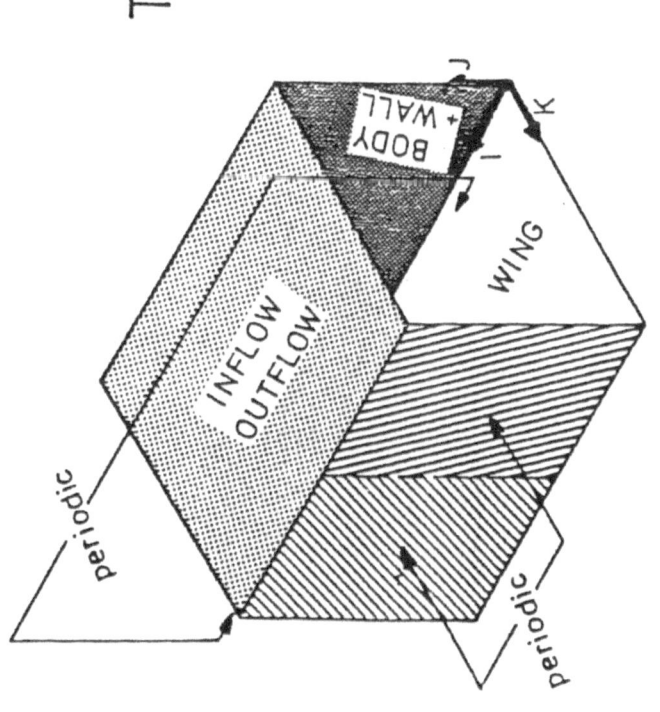

Fig. 5   The boundaries of the physical domain are mapped by the mesh to the six faces of the computational cube. The appropriate flow conditions must be set on these faces.

CONTIGUOUS CELLS IN MEMORY

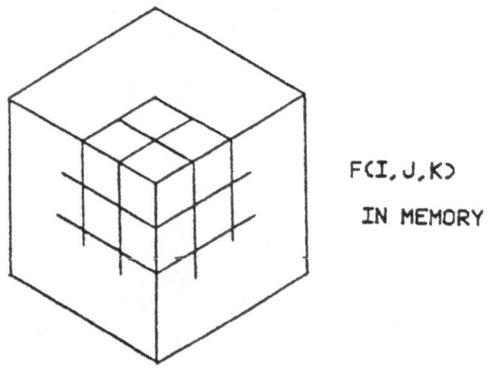

F(I,J,K)

IN MEMORY

EXPLODED VIEW

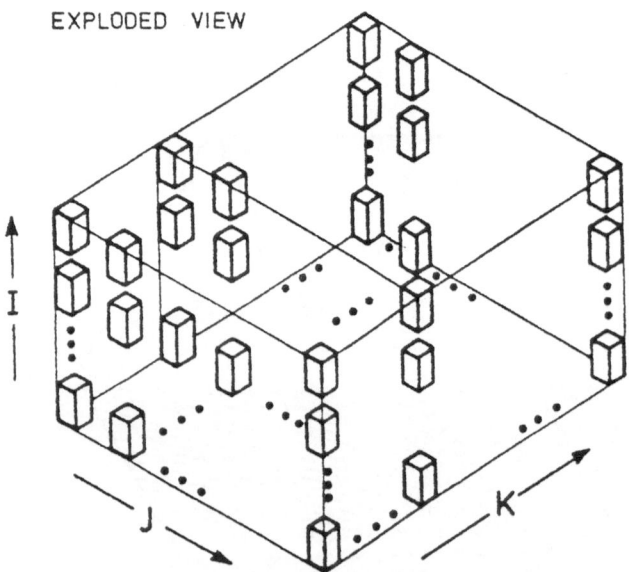

Fig. 6   Exploded view of the three-dimensional FORTRAN
          array F(I,J,K) for the flux illustrates the pencil
          concept for contiguons memory locations.

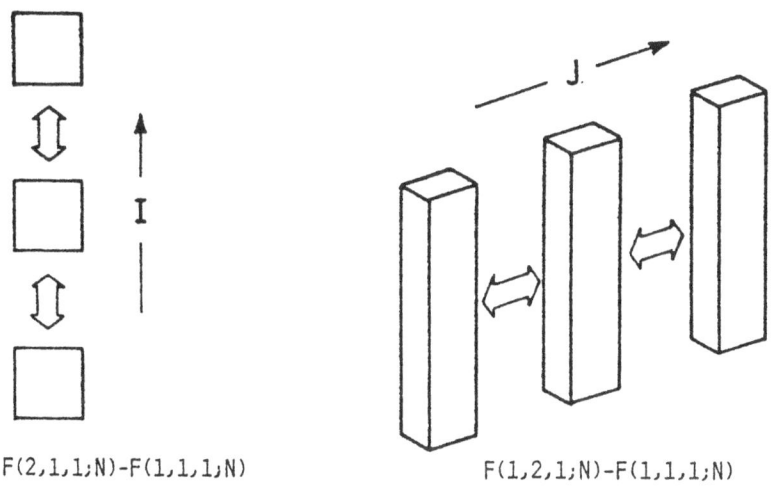

F(2,1,1;N)-F(1,1,1;N)   F(1,2,1;N)-F(1,1,1;N)

F(1,1,2;N)-F(1,1,1;N)

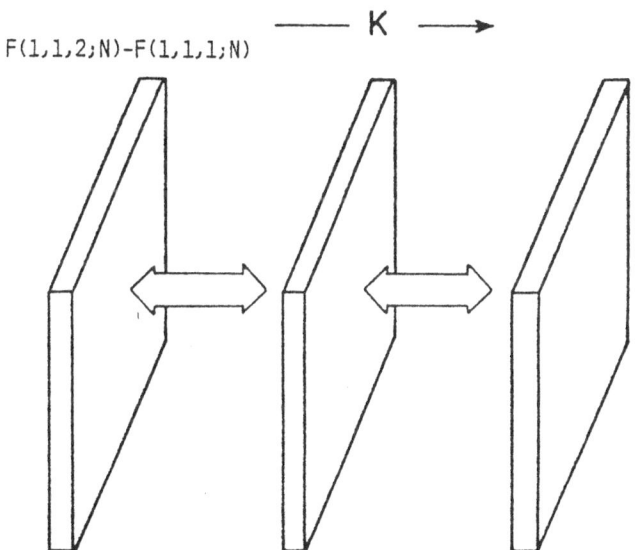

Fig. 7   Because of the grid structure the flux F can be differenced as a vector by offsetting its starting location.

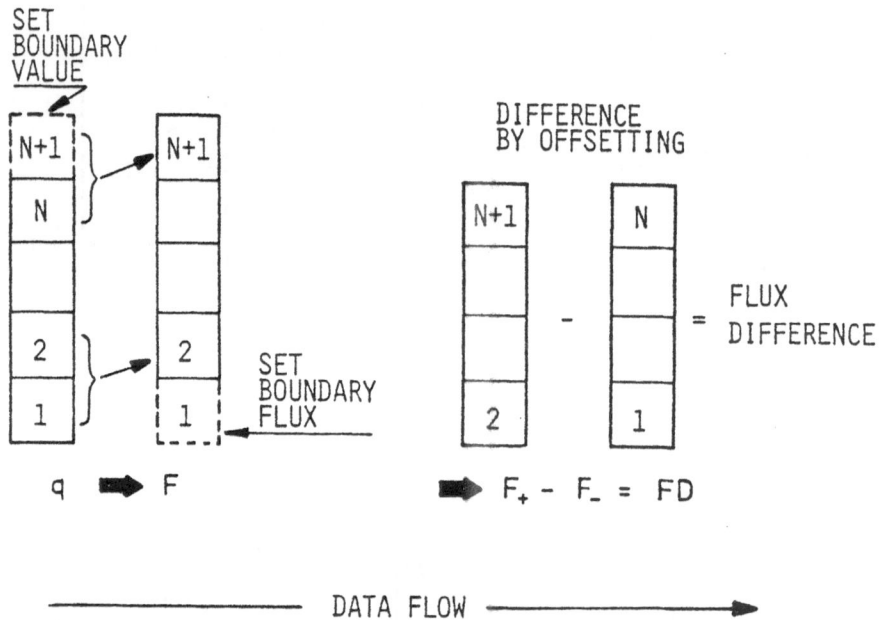

Fig. 8 Vector alignment and data flow in the algorithm
for interleaving the boundary conditions into
the vector differencing of the entire three-
dimensional flux field giving the total flux
differences.

STRIP MINING THE FLUX DIFFERENCING

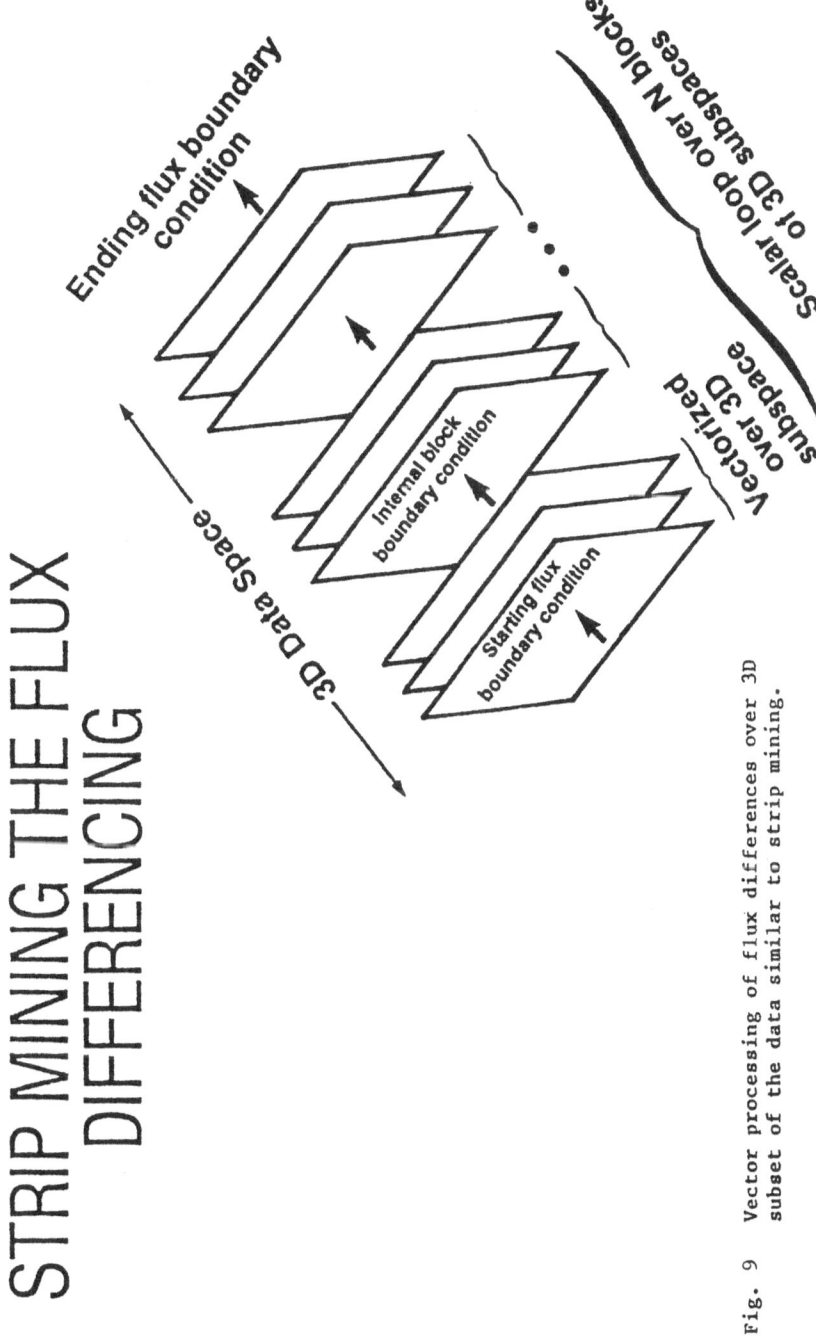

Fig. 9 Vector processing of flux differences over 3D subset of the data similar to strip mining.

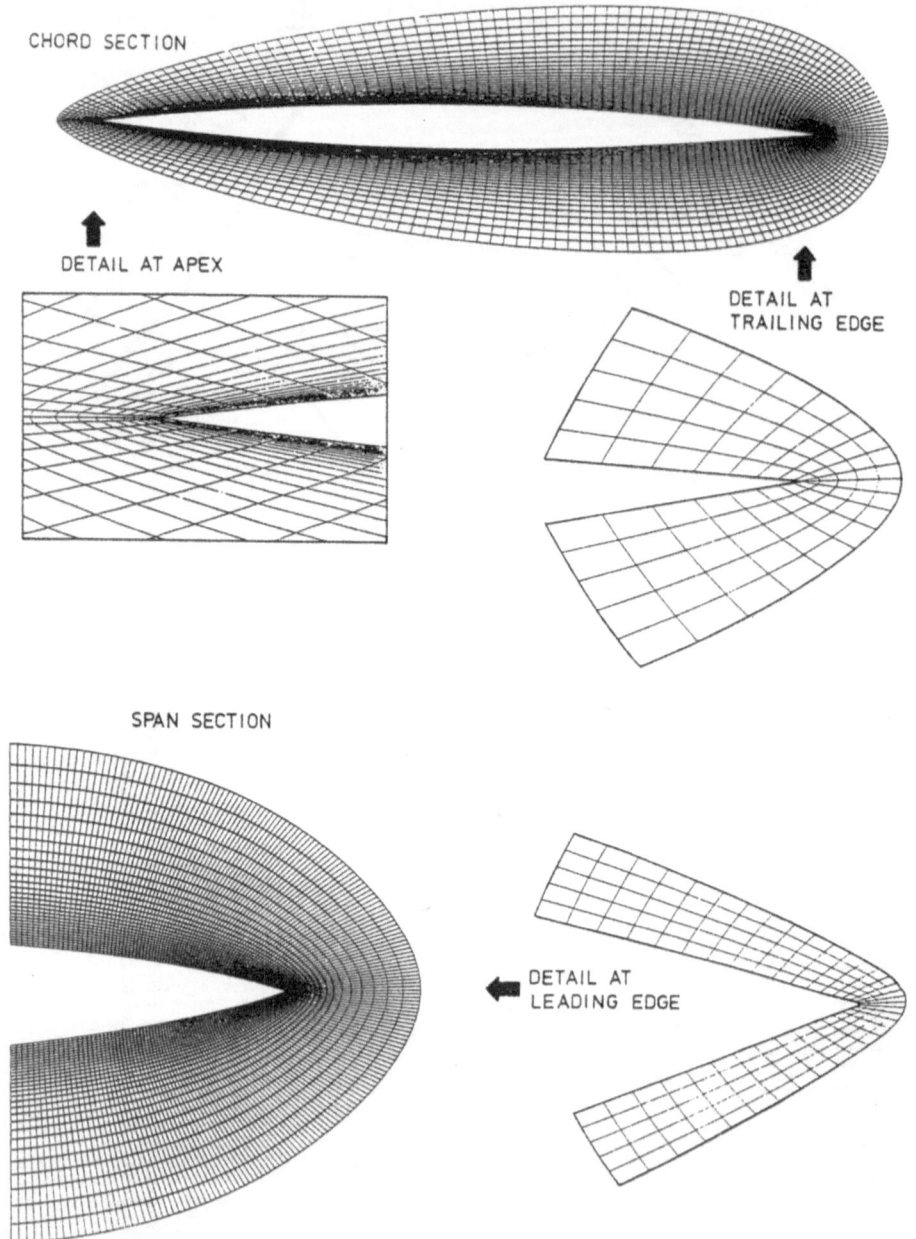

Fig. 10 Partical chordwise and spanwise views of
the 193x57x97 mesh.

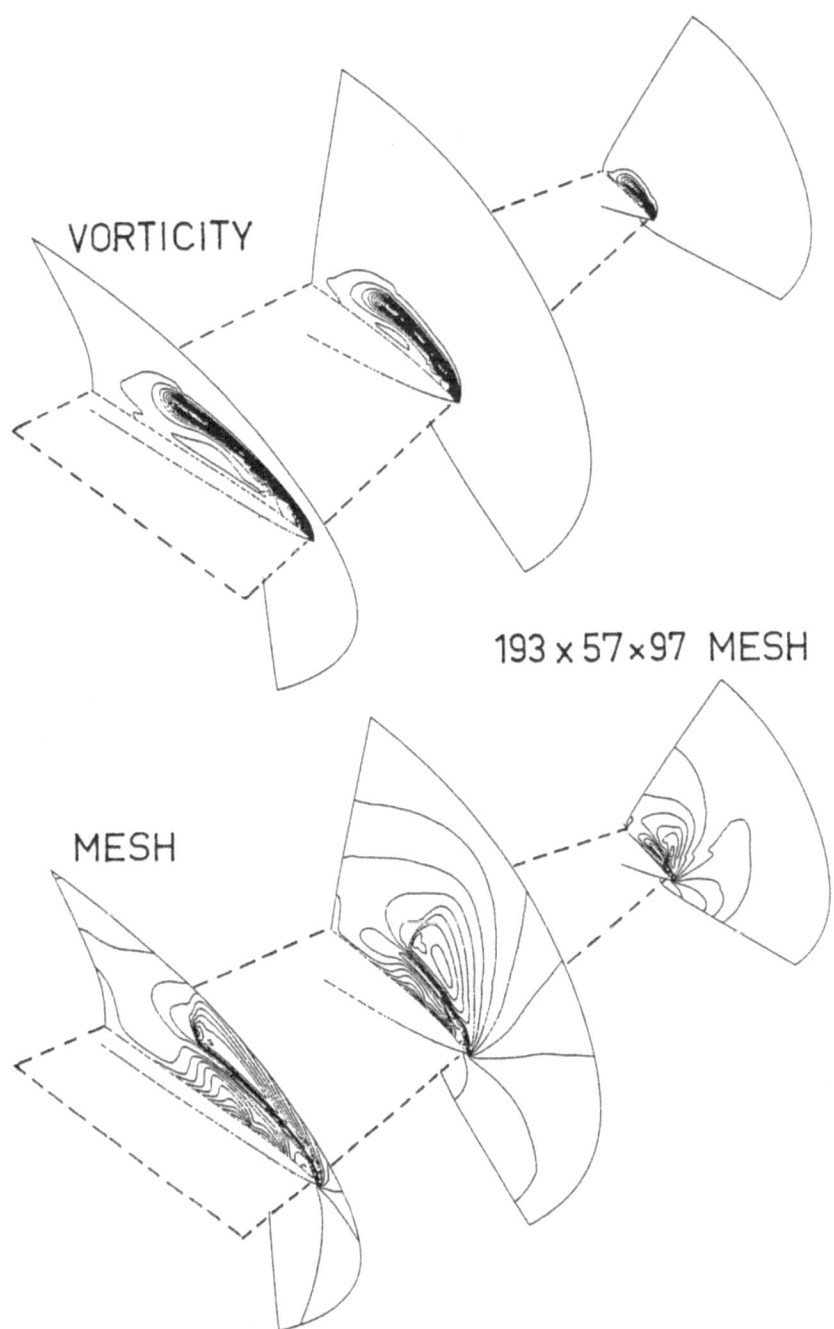

VORTICITY

193 x 57 x 97 MESH

MESH

Fig. 11   Three-dimensional isograms shown in parallel projec-
tion show the leading shock waves and vortical flow
over the upper surface. Dillner delta-wing $M_\infty = 1.75$
$\alpha = 18$ deg.   197x57x97 mesh.

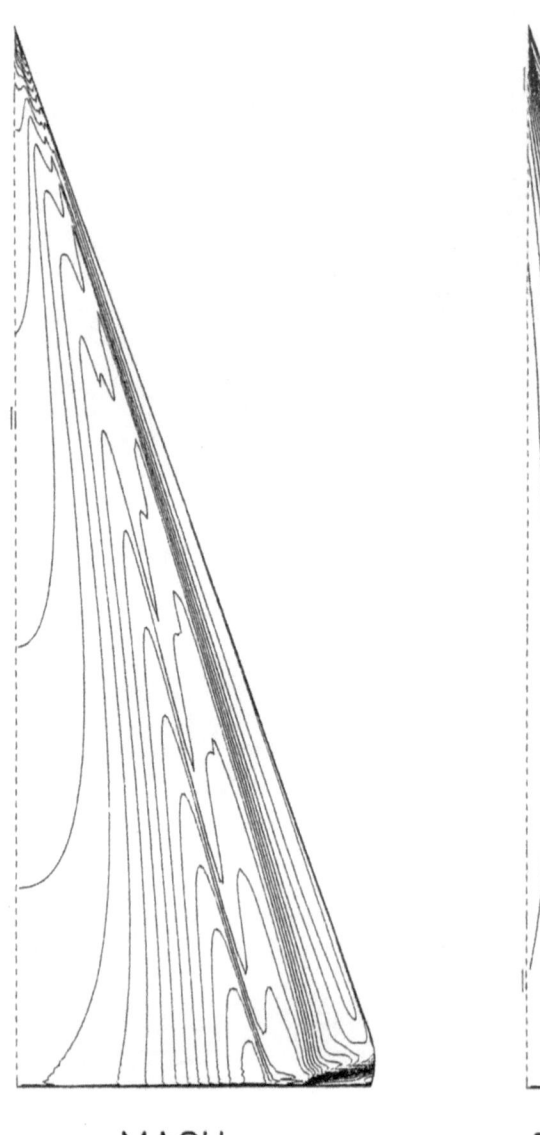 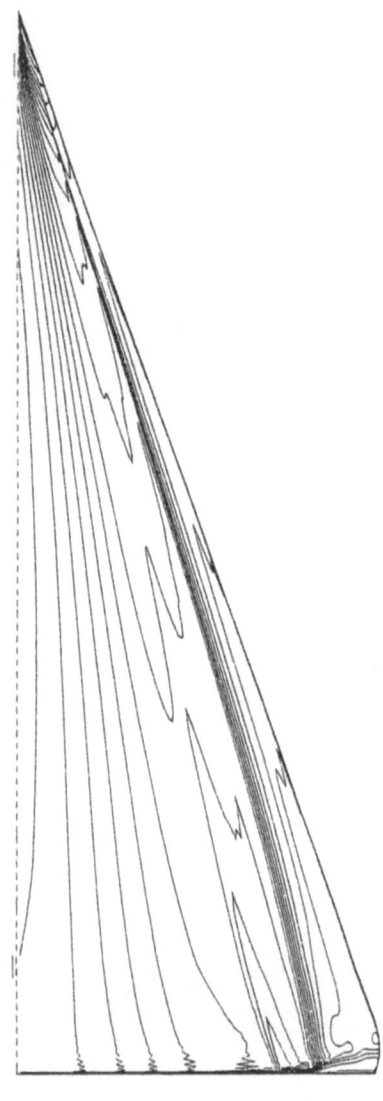

MACH   TOTAL PRESSURE

Fig. 12   Iso Mach and constant total pressure contours sho-
          wing the complex wave interactions on the upper
          surface of the wing.

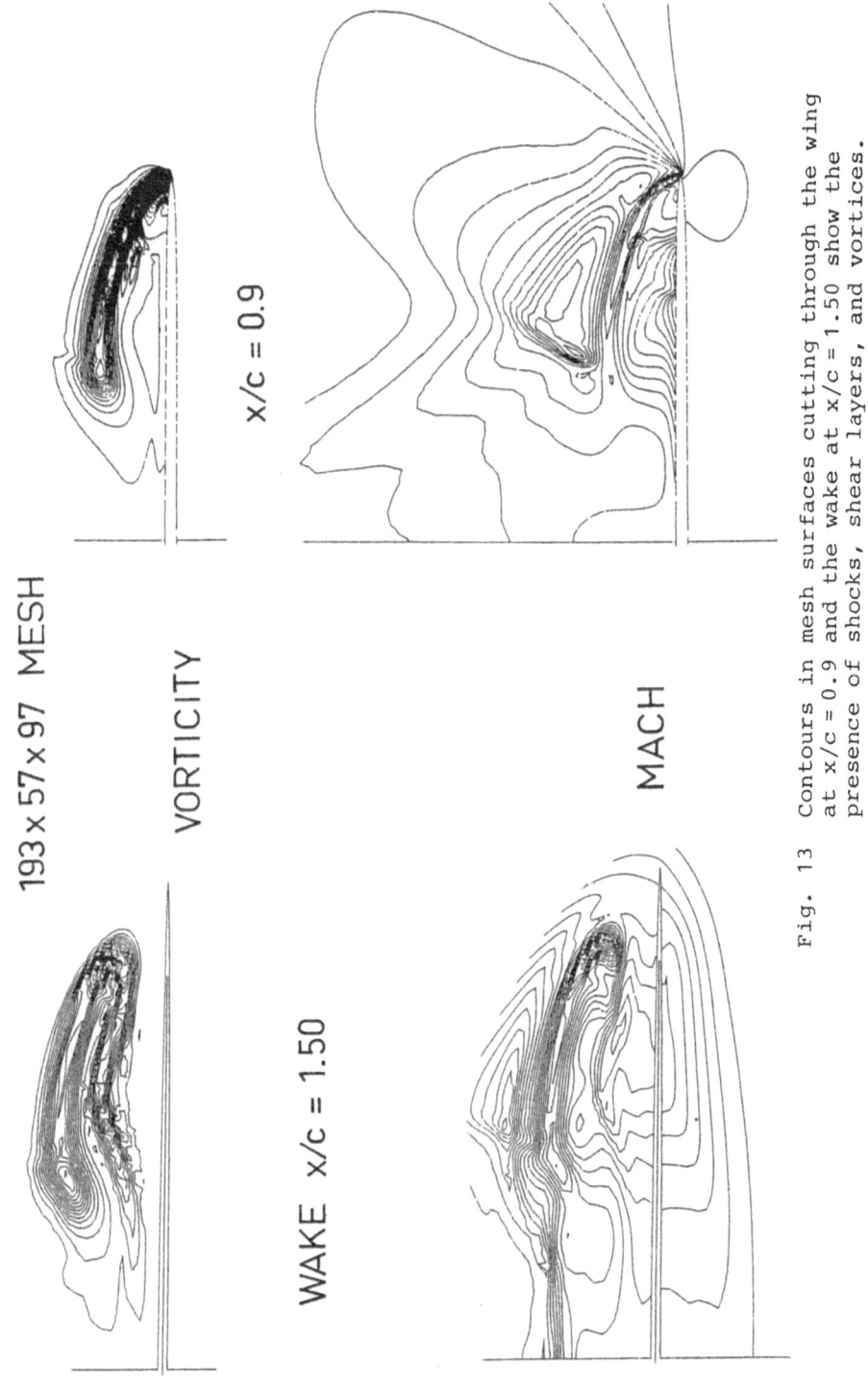

193×57×97 MESH

VORTICITY

x/c = 0.9

WAKE x/c = 1.50

MACH

Fig. 13   Contours in mesh surfaces cutting through the wing
at x/c = 0.9 and the wake at x/c = 1.50 show the
presence of shocks, shear layers, and vortices.

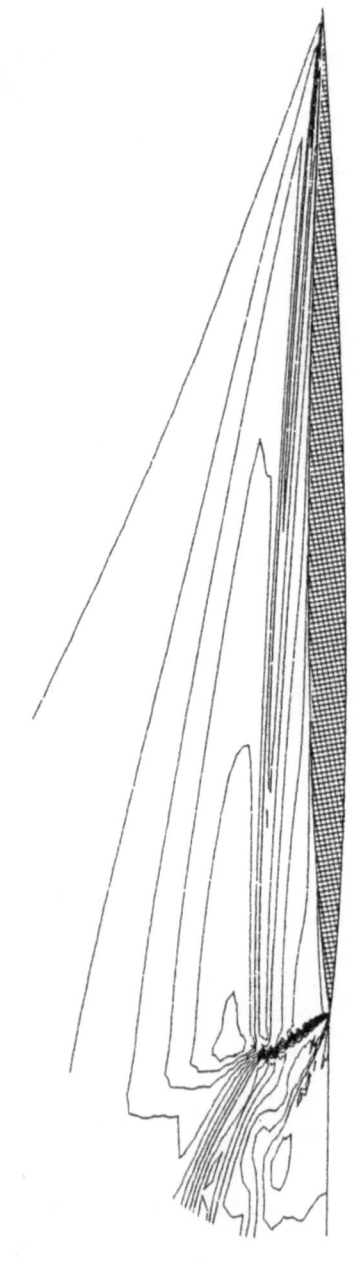

Fig. 14   IsoMach contours drawn axially through the vortex core
show the interaction of the vortex with the slipstream
and shock wave at the trailing edge. $M_\infty = 1.75$, $\alpha = 18$ deg.

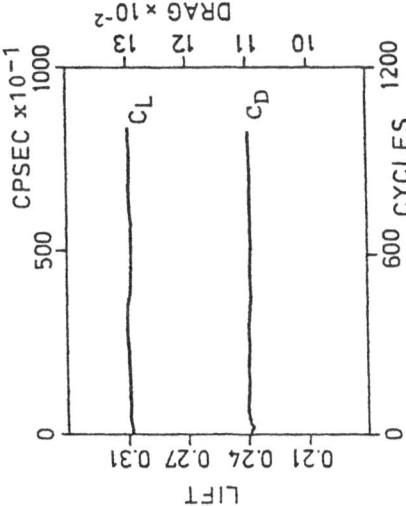

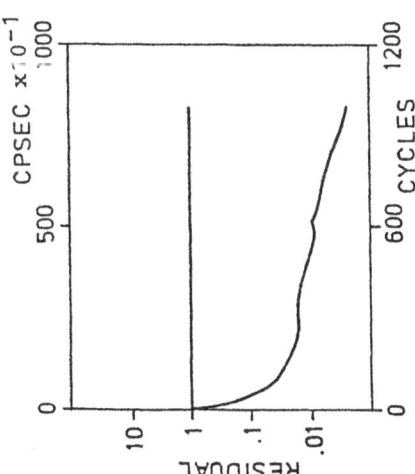

Fig. 15   Convergence history versus number of cycles and CPU
seconds on the CYBER 205. Freestream initial conditions.

COMPUTER PERFORMANCE

CYBER 205 16 M-WORDS REAL MEMORY

MESH	ONE MILLION POINTS (193x57x97)
DATA SET	23 M-WORDS (32 BIT)
INTERATIONS	1000
CPU TIME	2.5 HOURS
ELAPSED TIME	2.5 HOURS
UNIT TIME	8 MICROSEC PER GRID POINT PER ITER.
RATE	125 M FLOPS
TOTAL OPERATIONS	ONE TRILLION

Fig. 16 Performance of the 16 M-word CYBER 205 vector processor for the one-million-grid-point simulation in Fig. 15.

EFFECTIVE PROGRAMMING OF FINTE ELEMENT METHODS FOR COMPUTATIONAL FLUID DYNAMICS ON SUPERCOMPUTERS

R. Löhner, K. Morgan and O.C. Zienkiewicz

Institute for Numerical Methods in Engineering
University College of Swansea
Singleton Park, Swansea SA2 8PP, Wales, U.K.

SUMMARY

The effective programming of Finite Element Methods for CFD on vector-machines is discussed. It is shown, that for unstructured grids the performance observed on this class of machine depends heavily of the availability of hardware GATHER/SCATTER. Timings obtained for a 3-D Euler code are presented for the CYBER-205, CRAY-XMP11 and CRAY-XMP48.

INTRODUCTION

Our aim is to solve large, 3-D industrial high speed flow problems employing unstructured grids in conjunction with finite element methods. Other discretization procedures, such as finite difference or finite volume methods have reached a high degree of sophistication [1-4] , whereas finite element methods have not been applied on a large scale so far. However, due to their inherent geometrical flexibility, which is important when modelling industrial flow problems, they are expected to become a major force in CFD during the next few years. In what follows, we describe the basic solution algorithm employed, and the steps taken to achieve acceptable speed-up ratios on machines like the CYBER-205, CRAY-XMP11 and CRAY-XMP48.

ALGORITHMS

As from the onset we only consider <u>unstructured grids</u>, no 'optimal implicit algorithm' (such as ADI) can be envisaged, so it is apparent that explicit schemes will have to be employed. Explicit schemes imply : low storage (especially for unstructured grids, where implicit schemes require the inversion of large matrices), easy programming, easy vectorization, excellent time-accuracy for transient problems, in some cases slow convergence for steady-state problems (in this latter case unstructured multigrid-methods [16] are employed in conjunction with explicit schemes to accelerate the convergence).

The equations governing compressible fluid flow may be written in conservation form as

$$\frac{\partial \underline{u}}{\partial t} + \frac{\partial \underline{F}^i}{\partial x^i} = \underline{0}. \tag{1}$$

This equation is first discretized in time in a Lax-Wendroff manner [5-7], resulting in

$$\Delta \underline{u} = -\Delta t \frac{\partial \underline{F}^i}{\partial x^i} + \frac{\Delta t^2}{2} \cdot \frac{\partial}{\partial x^j} (\underline{\underline{A}}^j \frac{\partial \underline{F}^i}{\partial x^i}), \quad \underline{\underline{A}}^i = \frac{\partial \underline{F}^i}{\partial \underline{u}}, \tag{2}$$

and is then discretized in space by a straightforward Galerkin procedure. The justification for employing the same polynomials as weighting and test functions may be found in [6,7].

An undesirable feature in eqn.(2) is the appearance of the Jacobian matrices $\underline{\underline{A}}^i$, the construction of which requires excessive CPU-time (in particular for 3-D we have three 5*5 matrices that need to be evaluated per element). However, these Jacobians may be avoided by building the right-hand side of eqn(2) in a two-step procedure as follows :

$$\underline{u}^{n+\frac{1}{2}} = \underline{u}^n - \frac{\Delta t}{2} \cdot \left. \frac{\partial \underline{F}^i}{\partial x^i} \right|^n, \tag{3a}$$

$$\Delta \underline{u} = -\Delta t. \left. \frac{\partial \underline{F}^i}{\partial x^i} \right|^{n+\frac{1}{2}}. \tag{3b}$$

If one now develops the last equation in terms of a Taylor-series, the analytical equivalence between (3) and (2) is readily seen. However, the numerical equivalence is only achieved if constant shape functions are employed at the half step n+1/2.

For strong shocks an invariant form of the artificial viscosity due to Lapidus [8] is used [9].

GENERAL CONSIDERATIONS FOR SPEED-UP ON VECTOR MACHINES

In figure 1 the relative speed-up (speed/maximum possible speed) achieved by performing a certain percentage of the total amount of operations in vector-mode is depicted. As one can see, for machines with vector/scalar speed-up ratios of 1:20 (e.g. CYBER 205) we need to perform at least 95% of all operations in vector mode to achieve half the possible performance. This in turn implies that at least 95% of all operations must be performed inside "long" DO-loops (>20 for the CRAYs, >200 for the CYBER 205). In typical FEM programs, the contribution of each element is evaluated one-by-one (by calling the respective element subroutine). This implies that

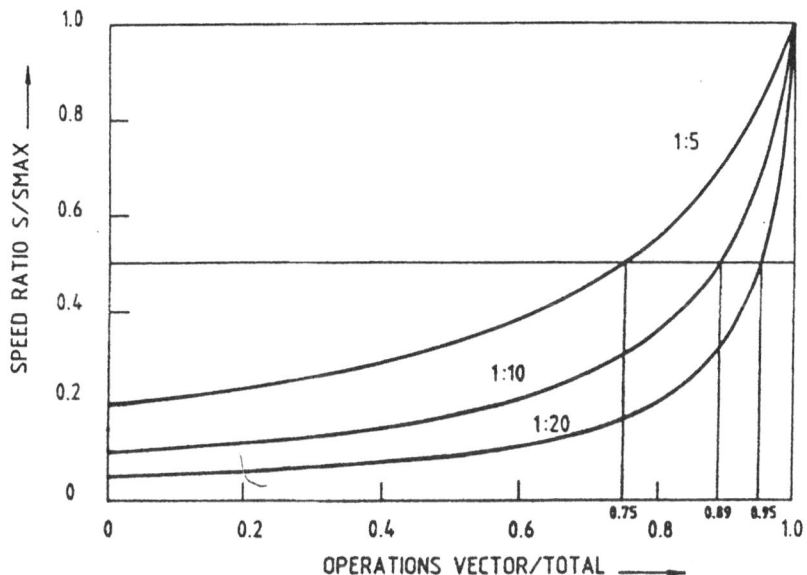

Figure 1   Expected performance achieved by vectorization for
computers with varying vector/scalar speed-up
ratios

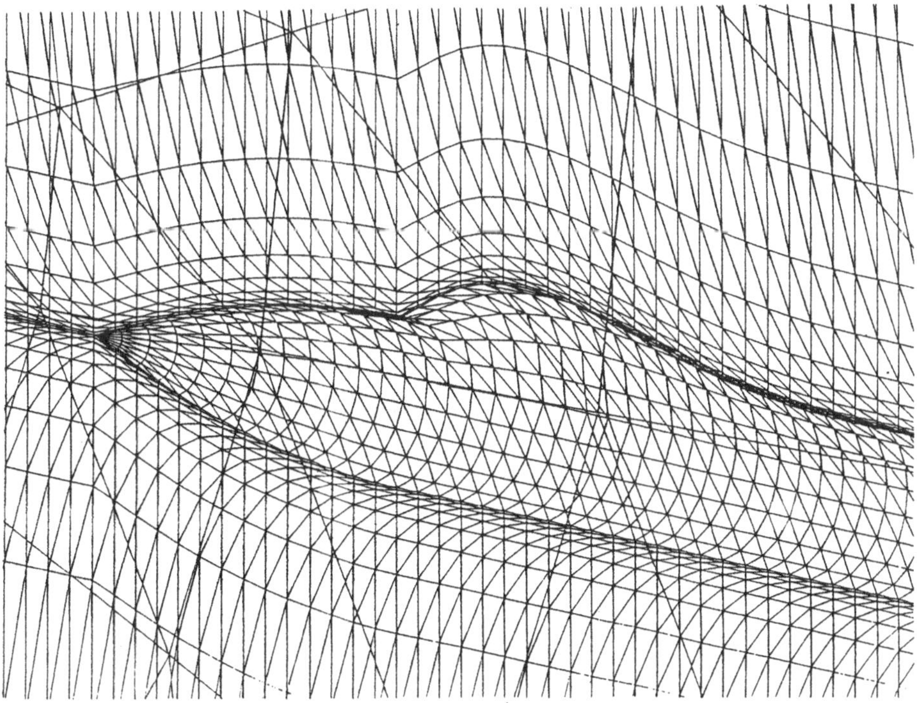

Figure 2   Detail of a typical 3-D FEM discretization
(57350 elements, 11350 nodes)

the typical DO-loop length is way too short (usually 4-8).
Therefore, this concept has to be abandoned, and the
contributions of many elements (possibly all) have to computed
at the same time. This involves a total restructuring of the
program, and not just the optimization of the most
time-consuming routines.

Moreover, in order to avoid the excessive storage
requirements of too many dummy-arrays, one-point integration or
the exact (analytical) evaluations of all integrals is
favoured. For this reason, we use simple elements (triangles
in 2-D, tetrahedra in 3-D), and, at present, "one-element"
programs (typical FEM programs allow the concurrent use of a
variety of elements). In this way, we never need to form a
matrix, which even on a scalar machine reduces CPU-times
considerably.

STORAGE

At present, we store the derivatives of the shape-functions
and the jacobians of the elements, although these could be
recomputed at each time-step, decreasing storage requirements
but increasing CPU-time. We are not sure at present which
strategy should be preferred.

A problem encountered when trying to maximize the speed is
the use of too many dummy-arrays, which require an excessive
amount of storage. The solution adopted here is to use
subdomains or blocks of elements, so that the dummy-arrays only
extend over a "small" number of storage locations. Each block
consists of approximately 6000 elements, so that acceptable
vector-lengths are still mantained. During each
timestep/iteration the whole data-base is scanned twice (this
can be reduced to one if the artificial viscosity is applied
inside the timestep and not after). The subdivision is
performed automatically by the program, and does not depend on
any ordering or topology of the mesh. In this way, it was
possible to halve storage requirements without sacrificing
speed.

Nevertheless, experience on the CYBER 205 indicates that
once the machine starts paging, the degradation in performance
is considerable. The CPU-speed/storage-fetch-speed ratio is
totally unbalanced (this is not the case for e.g. a VAX or
a PRIME).

USE OF SPECIAL ROUTINES

As the grids employed are unstructured, at least two
independent sets of data have to be processed and related to
each other during the computation : these are point-arrays
(unknowns/fluxes at points, right-hand sides, etc.), and
element-arrays (derivatives of shape-functions, jacobians,
etc.).

120

The first step of the algorithm (eqn.(3a)) consists mainly of a GATHER operation from points into elements with subsequent multiplication by shape-function derivatives. It presents no further difficulties.

The second step (eqn.(3b)) consists mainly in a SCATTER-ADD operation : the element contributions are added to the nodes. Being a semi-recursive operation (not every point obtains contributions from the same number of elements), it is not so easy to vectorize. Typically, it would be programmed as follows:

```
      DO 200 IP=1,NPOIN
      VP(IP)=0.0
200 CONTINUE                                         (4)
      DO 210 IE=1,NELEM
      VP(LE(IE))=VP(LE(IE))+VE(IE)
210 CONTINUE .
```

The simple GATHER-SCATTER routines are of no help here, so that 'in-house' special routines have to be written. For (4) we have chosen two alternatives :

a) Unwrapping/unrolling the DO 210 -loop into sets of 25 elements at a time has proved quite effective on the CYBER-205, but is, of course of limited use as it does not allow any further parallelism (this if of great importance for the future).

b) Reshuffle and subsequent piecewise scatter :

```
      DO 400 IE=1,NELEM
      L2E(IE)=LE(L1E(IE))              GATHER
400 CONTINUE
----
      DO 410 IE=1,NELEM
      TE(IE)=VE(L1E(IE))              GATHER
410 CONTINUE
      NELE1=0
      DO 420 IPASS=1,NPASS
       DO 430 IP=1,NPOIN
       TP(IP)=0.0
430  CONTINUE
       NELEO=NELE1+1                                 (5)
       NELE1=LPASS(IPASS)
       DO 440 IE=NELEO,NELE1
       TP(L2E(IE))=TE(IE)             SCATTER
440  CONTINUE
       DO 450 IP=1,NPOIN
       VP(IP)=VP(IP)+TP(IP)
450  CONTINUE
420 CONTINUE
```

In this case, four new arrays (L1E,L2E,TE,TP) have had to be introduced, but the 1:1 structure has been recovered, by constructing L1E and L2E in an appropiate way. For more details, the reader is referred to [10,11].
The operation count goes as follows :
a) Unwrapping :
   NOPER = NELEM + NPOIN   (non-vectorizable)
b) Reshuffle (eqn.(5)) :
   NOPER = 2*NELEM + 2*NPASS*NPOIN   (vectorizable)
Of course, a CDC- or CRAY-written SCATTER-ADD would be the best solution here.

TIMINGS

Typical timings (CPU-seconds/grid-point/time-step*10^5) obtained for the 3-D-Euler code are presented for the following set of machines (for a typical example, see figure 2):

a) ICL-2966 (VME operating system) at Swansea,
b) CYBER-205 (two-pipe, full precision) at UMRCC Manchester,
c) CRAY-XMP11 (no hardware GATHER/SCATTER) at Bracknell,
d) CRAY-XMP48 (hardware GATHER/SCATTER, one CPU running) at Mendota Heights, Minnesota;

	ICL2966	CY205	XMP11	XMP48
eqn.4, vectorizer off	2210.2	202.2	136.0	
eqn.4, vectorizer on		29.3	45.1	23.19
eqn.5, vectorizer off	2783.1			
eqn.5, vectorizer on		12.6	52.0	12.14

As one can see, the effect of the hardware GATHER/SCATTER is indeed impressive. The CRAY-XMP-11 with no hardware GATHER/SCATTER just emulates the scalar machine performance, and CPU times increase due to the additional number of operations needed for the vectorized SCATTER-ADD process. On the machines with hardware GATHER/SCATTER CPU-times are halved by using the vectorized SCATTER-ADD.

FURTHER IMPROVEMENTS IN EFFICIENCY

The inherent geometrical flexibility of unstructured grid may be exploited futher to reduce CPU and storage requirements. We here briefly report of three possibilities, which we have explored:

Domain splitting

As we employ explicit schemes, the allowable time-step is governed by the smallest element. In order to achieve maximum economy in CPU-time, without loss of time-accuracy, on grids with large variation in element size, we introduced a domain splitting technique. It allows us to advance the solution with different timesteps in different regions of the mesh in a time-accurate manner [12].

Adaptive refinement

In order to reduce the overall number of gridpoints in the mesh (and this means reducing storage and CPU-time), adaptive refinement techniques were introduced in [13-15]. These can either be mesh movement or mesh enrichment. We are currently favouring mesh enrichment for steady state problems, as the initial guess for the enriched mesh is obtained in a very economical way, and the subsequent CPU-time is spent in obtaining 'desirable details' at the locations where they are needed.

Unstructured Multigrid Methods

It is well known that multigrid-based techniques presently offer the fastest way to obtain the steady-state solutions of stationary problems. In the case of unstructured grids, the traditional nestedness of finer grids embedded in coarser ones has to be abandoned, resulting in unstructured multigrid methods. Preliminary results [16] have so far been very encouraging, and we are still exploring the use of multigrid techniques for both inviscid and viscous high speed flows.

As a general remark, one can say that all REAL-operations involved in domain-splitting, adaptive refinement and unstructured multigrid methods may be vectorized, whereas the complicated logistics of these algorithms (INTEGER-arrays) elude large-scale vectorization (however, it is found that these INTEGER operations only require a small percentage of CPU-time).

CONCLUSIONS

It has been demontrated, that efficient FEM-based algorithms can be obtained for vector machines. The performance degrades seriously if no efficient GATHER/SCATTER routines exist on the particular machine. The traditional concept of the "general FEM-program" had to be abandoned in favour of the one-simple-element program in order to achieve the gain in speed.

In this way, fully vectorized programs were obtained whilst retaining the flexibility of totally unstructured grids.

123

ACKNOWLEDGEMENTS

The authors would like to thank the Aerothermal Loads Branch of the NASA Langley Research Establishment for supporting this research under Grant NAGW-478.

REFERENCES

1. WOODWARD, P., COLELLA, P.: "The Numerical Solution of 2D Fluid Flow with Strong Shocks", J.Comp.Phys.54,115-173 (1984).

2. SMITH, R.E.: "Two Boundary Grid Generation for the Solution of the 3-D Compressible Navier-Stokes Equations", NASA TM 83123(1981).

3. SPRADLEY, L.W., STALNAKER, J.F., RATLIFF, A.W.: " Solution of Three-Dimensional Navier-Stokes Equations on a Vector Processor", AIAA J.19(10),1302-1308(1980).

4. BOOK, D.L.(edr.): "Finite Difference Techniques for Vectorized Fluid Dynamics Calculations", Springer Series in Computational Physics (1981).

5. ROACHE, P.:"Computational Fluid Dynamics",Hermosa Publishers (1970).

6. DONEA, J.:"A Taylor-Galerkin Method for Convective Transport Problems",Int.J.Num.Meth.Eng.20,101-119(1984).

7. LÖHNER, R., MORGAN, K., ZIENKIEWICZ, O.C.: "The Solution of Non-Linear Hyperbolic Equation Systems by the Finite Element Method",Int.J.Num.Meth.Fluids 4,1043-1063(1984).

8. LAPIDUS, A.: "A Detached Shock Calculation by Second-Order Finite Differences",J.Comp.Phys.2,154-177(1963).

9. LÖHNER, R., MORGAN, K., PERAIRE, J.:"A Simple Extension to Multidimensional Problems of the Artificial Viscosity Model due to Lapidus",to appear in Comm.Appl.Num.Meth.(1985).

10. DIEKKÄMPER, R.: "Vectorized Finite Element Analysis of Nonlinear Problems in Structural Dynamics", Proc. "Parallel Computing 83" Conf. (M. Feilmeier, G. Joubert and U. Schendel eds.),293-298, North-Holland (1984).

11. LÖHNER, R. , MORGAN, K.: "Finite Element Methods on Supercomputers : The SCATTER-Problem",Proc. NUMETA'85 Conf. (J. Middleton and G. Pande eds.), 987-990, A.A. Balkeema (1985).

12. LÖHNER, R., MORGAN, K., ZIENKIEWICZ, O.C.: "The Use of Domain Splitting with an Explicit Hyperbolic Solver",Comp. Meth.Appl.Mech.Eng.45,313-329(1984).

13. LÖHNER, R., MORGAN, K. ZIENKIEWICZ, O.C.: "An Adaptive Finite Element Method for High Speed Compressible Flow, Lecture Notes in Physics 218, pp 388-392, Springer (1985).

14. LÖHNER, R., MORGAN, K., ZIENKIEWICZ, O.C.: "An Adaptive Finite Element Procedure for Compressible High Speed Flows, Comp.Meth.Appl.Mech.Eng.(1985). To appear.

15. LÖHNER, R., MORGAN, K., ZIENKIEWICZ, O.C.: "Adaptive Grid Refinement for the Compressible Euler Equations", chapter in "Accuracy Estimates and Adaptivity for Finite Elements", Wiley, Chichester, 1985 (to appear).

16. LÖHNER, R., MORGAN, K.: "Unstructured Multigrid Methods: First Experiences",in Proc. Conf. on Num.Meth. in Thermal Problems, Pineridge Press, Swansea 1985.

PERFORMANCE EVALUATION OF EXPLICIT SHALLOW-WATER EQUATIONS SOLVERS ON THE CYBER 205

F.W. Wubs

Centre for Mathematics and Computer Science (CWI)
Kruislaan 413, 1098 SJ Amsterdam, The Netherlands

SUMMARY

The performance of an explicit method and an ADI method for the shallow-water equations is compared on a CYBER 205. Furthermore, a stabilization technique is discussed, which stabilizes the explicit method in such a way that any desired time step is possible without the development of instabilities.
Comparing the codes for two test models, we found that the explicit methods are attractive on the CYBER 205.
Finally, some proposals are made for the handling of irregular geometries.

INTRODUCTION

Designing a solver for the hyperbolic shallow-water equations for use on a CYBER 205, we would like to have some insight in the performance of explicit and implicit methods on such a computer. Therefore, we implemented an explicit method on the CYBER 205 and compared the performance of this method with an existing ADI method. However, the available code of the ADI method was not vectorized. Currently this method is in the final stage of vectorization and the computation speed is already known.
For some problems the time step restriction of an explicit method is severe. In order to overcome this restriction, we constructed an explicit stabilization technique, which allows us to use any desired time step.

THE SHALLOW-WATER EQUATIONS

In hydraulic engineering, the shallow-water equations (SWE's) are used to describe flows in shallow seas, estuaries and rivers. Output from the numerical models based on these SWE's can be used as input for models describing the influence of infrastructural works, salt intrusion, the effect of waste discharges, the water quality, cooling water recirculation and sediment transports. An important application, in the Netherlands, is the storm surge barrier under construction in the mouth of the Oosterschelde Estuary, by which this estuary can be closed off during storms. In this case, the numerical model, based on the SWE's, provides guide lines for the operation of the barrier in order to preserve the delicate ecological balance in the estuary, which has an important fish nursery as well as oyster and mussel cultures. Therefore, the construction of efficient numerical models for the SWE's is relevant.

The nature of the applications is such that strong gradients are common, though shocks do not appear. As a consequence, it is not neccesary to satisfy numerical conservation of momentum or energy. However, the conservation of mass is important as the depth determines largely the prolongation of the waves.

In Figure 1, we have drawn a cross section of a sea.

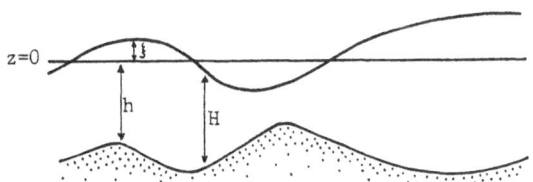

Figure 1 Schematisation of a cross section of the sea.

Having a reference plane z=0, which is for example the mean level of the sea, we define the local bottom profile by h(x,y) and the local elevation by ξ(x,y,t). Now, the total depth is given by $H=h+\xi$. Furthermore, averaging the velocities over the depth and assuming hydrostatic pressure and incompressibility of water, the SWE's read:

$$u_t = -uu_x - vu_y - g\xi_x + fv - Cz\sqrt{u^2+v^2}\ u/H + \upsilon\Delta u,$$
$$v_t = -uv_x - vv_y - g\xi_y - fu - Cz\sqrt{u^2+v^2}\ v/H + \upsilon\Delta v, \tag{1}$$
$$\xi_t = -(Hu)_x - (Hv)_y.$$

The first two equations are momentum equations describing, in this incompressible case, the change in time of the averaged velocities u and v. The third one is a continuity equation. In the momentum equation appear the Coriolis force parametrized by f, which is due to the rotation of the earth, and the bottom friction parametrized by Cz. Furthermore, g and υ denote the acceleration due to gravity and the diffusion coefficient for horizontal momentum, respectively. In practice, even more terms are introduced in the equations, but we have restricted ourselves to the most important of them. In practical numerical calculations the grid sizes are so large, of order 100m, that the significance of the diffusion terms is small with respect to both accuracy and stability. As a consequence, the equations behave numerically as hyperbolic equations.

EXPLICIT VERSUS IMPLICIT METHODS

On a vector computer, it is not a priori clear whether implicit or explicit methods should be used. Therefore we have collected some arguments to help us in that decision:

Table 1 Arguments pro and contra explicit and
         implicit methods.

EXPLICIT	IMPLICIT
simple implementation no recursion long vectors simple to decompose 　the domain	difficult implementation recursion short vectors(ADI) difficult to decompose 　the domain
time step restriction	no time step restriction

One could also take into account the need for workspace of the methods.
However, as the vector computer CYBER 205 uses dynamically storage, this is
not easy to determine. For example, a temporary variable, occurring in a
"do loop", to which an array element is assigned, transforms to an array of
the length of the "do loop" on a vector machine. Taking this into account
it is questionable whether an explicit method uses less storage than an
implicit method on a vector computer. But, as the SWE's are a 2-D problem,
the storage is not the main problem.
From the above arguments, we decided to use an explicit method. The
consequence of this choice is the restriction on the time step, which may
be severe for problems, slowly varying in time. Because, in that case we
would like to use a large time step. Fortunately, by using a stabilization
technique, we were able to overcome this drawback.

                              DISCRETIZATIONS

In this section, we will shortly describe the ADI method and the explicit
method we used in our comparison.

An ADI method

The method given here was designed by Stelling [4] in 1983. It is claimed
that this method has good stability properties with respect to the
advection terms, which are in general quite troublesome. Furthermore, on a
sequential computer the storage requirements are very low. For the tests we
performed, we had not a vectorized version of this program. Currently, the
program is in the final stage of vectorization. The main problem for the
vectorization is the solution process for the system of equations occurring
after discretization. Here, this is handled by using cyclic reduction which
vectorizes to a good degree. The execution time of the vectorized version
for long vectors is about 2. 10^{-5}seconds per grid point per time step on a
one-pipe CYBER 205 in half precision, whereas this number is 1.5 10^{-4} for
the scalar version.
For the introduction of the ADI method we use the method of lines
approach[3]. In this approach, first (1) is semi-discretized with respect
to space, giving the system of ordinary differential equations

128

$$\frac{d}{dt} \vec{W} = \vec{F}(\vec{W}), \tag{2}$$

where $\vec{W}=(\vec{U},\vec{V},\vec{\xi})$ and $\vec{F}(\vec{W})$ is a short form for the discretized right-hand side of (1). Then, for the time integration of (2), an appropriate time integrator is used. In this case, the time integrator is given in the so-called split notation [1]:

$$\vec{W}^{n+\frac{1}{2}} = \vec{W}^n + 1/2\Delta t \; \vec{G}(\vec{W}^{n+\frac{1}{2}},\vec{W}^n),$$
$$\vec{W}^{n+1} = \vec{W}^{n+\frac{1}{2}} + 1/2\Delta t \; \vec{G}(\vec{W}^{n+\frac{1}{2}},\vec{W}^{n+1}), \tag{3}$$

where the splitting function $\vec{G}$ satisfies $\vec{G}(\vec{W},\vec{W})=\vec{F}(\vec{W})$. It is straight-forward to show that (3) is second-order in time. Furthermore, the components of G are described by

$$\vec{G}_1(\vec{W},\overset{\backsim}{\vec{W}}) = -[\; U\overset{\backsim}{U}_x + V\overset{\backsim}{U}_y + g\xi_x \; \ldots], $$
$$\vec{G}_2(\vec{W},\overset{\backsim}{\vec{W}}) = -[\; \overset{\backsim}{U}V_x + \overset{\backsim}{V}V_y + g\overset{\backsim}{\xi}_y \; \ldots], \tag{4}$$
$$\vec{G}_3(\vec{W},\overset{\backsim}{\vec{W}}) = -[\; (UH)_x + (\overset{\backsim}{VH})_y \;],$$

where only the most important terms are given. The brackets [] denote the space discretization of the terms within it. This discretization, which is done on a space staggered grid, is second order and central for all terms except for the cross terms VU_y and UV_x which are treated third order. This third order treatment is chosen because of its damping properties for high frequency components in the numerical solution.

In the time discretization, the treatment of the advection terms is different from what is done usually in ADI methods. For example, in the first stage, the momentum equation for V is treated explicitly in the elevation (pressure) term and implicitly in the advection terms. Usually the advection terms are treated also explicitly in this stage. For a more detailed discussion of this discretization and its properties we refer to the author's thesis [4].

The method has the important property that the implicit systems contain only tridiagonal matrices, which allow a fast solution process.

An explicit method

The method we implemented is straight-forward. The time integration is given by:

$$\vec{W}^{n+1} = \vec{W}^n + \Delta t \; (\vec{K}_1 + 2\vec{K}_2 + 2\vec{K}_3 + \vec{K}_4)/6, \tag{5}$$

where

$$\vec{K}_1 = \vec{F}(\vec{W}^n),$$
$$\vec{K}_2 = \vec{F}(\vec{W}^n + 1/2\Delta t \; \vec{K}_1),$$
$$\vec{K}_3 = \vec{F}(\vec{W}^n + 1/2\Delta t \; \vec{K}_2), \tag{6}$$
$$\vec{K}_4 = \vec{F}(\vec{W}^n + \Delta t \; \vec{K}_3).$$

This is the classical fourth-order Runge-Kutta method [3]. The imaginary stability boundary of this method is $C=2\sqrt{2}$. The space-discretization is also treated fourth-order consistent. However, we have added a fourth-order diffusion term with an $O(\Delta x^3)$ term (Δx is the mesh size in space directions) in front of it, which has the same effect as a third order treatment of the advection terms. This term can be incorporated, without costs, into the physical diffusion, because this only involves a change of constants.

The execution time for long vectors for this method is 6.10^{-6} seconds per grid point per time step for half precision calculation on a CYBER 205, which is approximately three times faster than the vectorized ADI-implementation.

STABILIZATION OF THE TIME INTEGRATION

For some problems the stability condition of the explicit method is much to restrictive. Hence, we tried to stabilize the explicit method in order to be able to use larger step sizes. Here, we shortly give the basic idea. For details we refer to the reports [5] and [6].

Consider the hyperbolic equation

$$\vec{w}_t = \vec{f}(\vec{w},\vec{w}_x,\vec{w}_y,x,y,t), \qquad (x,y)\epsilon R^2, \ t>0, \tag{7}$$

where $\vec{w}=(\vec{u},\vec{v},\vec{\xi})$. Now, under certain conditions, we can show that, if the time derivatives of $\vec{w}$ are small then the space derivatives of $\vec{f}$ are also small, where $\vec{f}$, after substitution of the exact solution, is only a function of x,y and t.

This property means that the right-hand side function is quite smooth in the space directions if the solution varies slowly in time. Note that this does not imply that $\vec{w}$ itself is smooth in space directions. Hence, if the solution varies slowly in time, then this property of a smooth right-hand side allows us to smooth the right-hand side of the discretized equations. Using the method of lines approach[3], semi-discretization of (7) results in the system of ordinary differential equations

$$\frac{d}{dt}\vec{W} = \vec{F}(\vec{W}). \tag{8}$$

Instead of (8), we propose to solve

$$\frac{d}{dt}\vec{W} = S \ \vec{F}(\vec{W}), \tag{9}$$

where S is a smoothing operator satisfying

$$S = I + O(h^2). \tag{10}$$

We give here two examples of smoothing operators. First an implicit one:

$$- \mu(S\vec{F})_{j-1} + (1+2\mu) \ (S\vec{F})_j - \mu(S\vec{F})_{j+1} = F_j. \tag{11}$$

This operator is applied first in x-direction and thereafter in y-direction. The stability condition is now given by

$$\Delta t < C \ \Delta x\sqrt{\mu} \ 1/(7/6\sqrt{2 \ gH}), \tag{12}$$

where C is the imaginary stability boundary of the time integrator. The same operator is used by Jameson[2]. He uses it to stabilize explicit methods for boundary-value problems. However, using a vector computer, we

are more interested in explicit smoothing operators.
An explicit smoothing operator is given by

$$S := \prod_{k=1}^{n} S_k,$$ (13)

where

$$(S_k \vec{F})_j = \mu_k F_{j-2^{k-1}} + (1-2\mu_k)F_j + \mu_k F_{j+2^{k-1}}, \quad \mu_k < 1/4.$$ (14)

Again the operator is applied in x- and y-direction, successively. Using this operator, the stability condition is, for $\mu_k=1/4$,

$$\Delta t < C\Delta x \; 2^n \; 3\sqrt{3}/4 \; 1/(7/6\sqrt{2} \; gH).$$ (15)

In practical computations we choose μ_k smaller than 1/4 such that the constant $3\sqrt{3}/4$ can be replaced by 1. This choice assures that the smoothing operator (14) is diagonal dominant for all k, which appeared to be essential for initial-boundary-value problems. With both methods arbitrary step sizes are possible; in the implicit method μ should be chosen such that (12) is satisfied and in the explicit method n should be chosen such that (15) is satisfied.
The advantage of the second operator is that it vectorizes much better than the implicit operator. We found that one smoothing operator of the form (14) is 25 times faster than the implicit smoother in a half precision calculation. In this experiment, the decomposition of the implicit smoothing operator was performed beforehand as far as possible. As for our purpose an increase of the time step by a factor 8 or 16 (obtained by 3 and 4 smoothings respectively) is usually sufficient, the explicit smoothing is about 8 or 6 times as fast as implicit smoothing. This factor will become even larger if more vector pipes are used.
Furthermore, the cost of one explicit smoothing is about one nineth of the $\vec{F}$ evaluation (in case of the SWE's). As the $\vec{F}$ evaluations determine the costs of the whole integration, one smoothing increases also the costs of the whole integration with this factor.

RESULTS

We have tested both methods on two simple geometries: a one-dimensional problem and a two-dimensional symmetric problem. The tests were performed on a one-pipe CYBER 205 in half precision, which means a machine precision of approximately 7 digits.

A one-dimensional problem

The geometry of the first problem is drawn in Figure 2. It consists of a basin of length 50km with a bump in the middle.

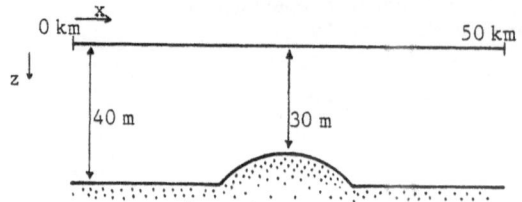

<u>Figure 2</u> Geometry of a one-dimensional problem.

The bottom profile is given by

$$h(x,y) = \begin{cases} 40 - 10 \sin((r+1)\pi/2) \ [m] \ \text{for} \ |r| < 1, \\ 40 \ [m] \qquad\qquad\qquad \text{elsewhere,} \end{cases} \tag{16}$$

where

$$r = (x-2.5 \ 10^4)/ \ 10^4. \tag{17}$$

At the boundaries the elevation is prescribed by

$$\xi(0,t) = - \sin(wt) \ [m], \tag{18}$$
$$\xi(50.10^3,t) = - \sin(wt-\phi) \ [m],$$

where

$$w = 2\pi/(12*3600) \ [s^{-1}], \tag{19}$$
$$\phi = 2\pi \ 5/60.$$

Furthermore, the constants in the equations (1) are chosen

$$f = 0 \ [s^{-1}],$$
$$\upsilon = 0 \ [m^2/s], \tag{20}$$
$$Cz = 4. \ 10^3.$$

We integrated 28 hours physically with various time steps (see Table 2). The calculations were performed on a 48x48 grid. At the end of each calculation we compared the solution with a reference solution computed on a finer grid (96x96). The results are given in significant digits of the ξ-component, defined by

$$Sdm = -^{10}\log \ (\ \max|\xi -\xi_{ref}| \ / \ \max|\xi_{ref} -\bar{\xi}_{ref}|) \tag{21}$$

where the subscript ref denotes the reference solution and the bar above $\bar{\xi}_{ref}$ denotes the average over the whole computational domain. The results are given in Table 2.

Table 2 Significant digits for
the one-dimensional problem

Δt[s]	ADI method			Explicit method		
	Computation time [s]		Sdm	Computation time [s]	Sdm	Number of smoothings
	scalar	vector				
21				78	2.3	0
42				39	2.3	0
84	454	65	1.5	22	2.1	1
168	227	32	1.5	13	1.5	2
336	114	16	1.5	8	.9	3
672	56	8	1.1			

In the first column, the time step is given; it increases downward with a factor two. In the second and fifth column the computation times are given. In the third column, we have added the expected computation times of the vectorized version of the ADI method. In the fourth and sixth column, the significant digits are given. In the last column, we have given the number of smoothings used in the product sequence (13). The time step of 42 seconds is close to the maximum time step allowed without smoothing.

With respect to accuracy, we see for both methods the same effect when the time step increases. At first, the number of significant digits changes slightly; then, when the time step is larger than 84 and 336 seconds for the explicit and implicit method, respectively, the number of significant digits decreases rapidly. This can be understood by the following reasoning. At first the error due to the space discretization, largely induced by the bump, is larger than the error due to the time discretization. This time discretization error is largely determined by the time-dependent boundary conditions (18). Then, when the time step increases the error due to the time discretization becomes more and more significant and even larger than the error due to space discretization.

With this in mind, we did not performe calculations with the implicit method for the time steps 21 and 42 seconds. The calculations would become rather expensive for these time steps, but the error would be the same as for the time step 84 seconds.

From the results we see that after vectorization of the ADI method the explicit method is generally faster even if a rather modest accuracy is required.

A symmetric two-dimensional problem

Now, we choose a geometry as drawn in Figure 3. In the middle of the basin, which has in this case sides of length 5km, we have placed a round bump.

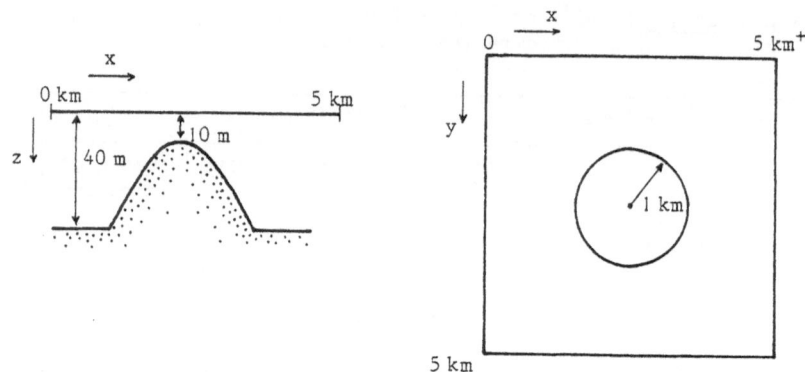

Figure 3 Geometry of the symmetric two-dimensional problem.

The bottom profile is in this case given by

$$h(x,y) = \begin{cases} 40 - 30\cos(\pi*r/2) \text{ [m] for } r<1, \\ 40 \text{ [m]} \qquad\qquad \text{ elsewhere,} \end{cases} \qquad (22)$$

where

$$r = \sqrt{(x - 2.5 \ 10^3)^2 + (y - 2.5 \ 10^3)^2} / 1000. \qquad (23)$$

At the left and right boundary, again the elevation is prescribed

$$\xi(0,y,t) = - \sin(wt), \qquad\qquad\qquad\qquad\qquad (24)$$
$$\xi(5.0 \ 10^3,y,t) = - \sin(wt-\phi), \quad 0<y<5.0 \ 10^3 \text{[m]},$$

where

$$w = 2\pi/(12*3600) \text{ [s}^{-1}], \qquad\qquad\qquad (25)$$
$$\phi = 2\pi \ 5/600.$$

At the upper and lower boundary, at $y=0$ and $y=5.0 \ 10^3$ [m] respectively, the normal velocity component v is zero. Furthermore, the constants in the equations (1) are chosen

$$f = 0 \text{ [s}^{-1}], $$
$$\upsilon = 10 \text{ [m}^2/\text{s]}, \qquad\qquad\qquad\qquad\qquad (26)$$
$$Cz = 4. \ 10^{-3}.$$

We integrated 15 hours physically with various time steps (see Table 3). The calculations were performed on a 24x24 grid. At the end of each calculation we compared the solution with a reference solution computed on a finer grid (96x96). The results are now given in significant digits of the v-component. In this case we have added also a root-mean-squares error given by

$$Sd2 = -^{10}\log (\ |v - v_{ref}|_2 / \ |v_{ref} - \bar{v}_{ref}|_2), \qquad (27)$$
$$|v|_2 = \sqrt{\sum_i v_i^2} \ /(24X24),$$

where the summation is over all grid points. The results are given in Table 3.

Table 3　Significant digits for the
symmetric two-dimensional problem

Δt[s]	ADI method				Explicit method			
	Computation time [s]		Sd2	Sdm	Computation time [s]	Sd2	Sdm	Number of smoothings
	scalar	vector						
8					45	2.1	1.3	0
16					26	2.1	1.3	1
32	181	40	1.6	1.1	15	2.1	1.3	2
64	91	20	1.4	.8	9	1.6	1.0	3
128	46	10	.8	.2				

Globally, we observe the same effect with the errors as in the previous
case. At first, the number of significant digits remains constant as the
time step increases and then, when the time step becomes larger than 32
seconds with both methods, the error due to the time step becomes dominant.
Again we see that to obtain a certain accuracy the explicit method is
faster than the ADI method.
On this 24X24 grid, we have lost already a factor 3 with respect to the
asymptotical speed on a very fine grid due to the start up times of the
vector instructions. The asymptotical speed is for the explicit method
almost reached for the 96X96 grid.
It is clear that without the stabilization technique the explicit method
would become too expensive with respect to a vectorized version of the ADI
method.

IRREGULAR GEOMETRIES

In the near future, we want to start calculations for irregular geometries.
When in such a case the computational domain is covered with a rectangle,
on which a fast computation can be performed, a lot of idle computations
on "land points" are done. To minimize this number of land points, we have
two proposals. The first is to decompose the domain and the second is the
use of two numberings.

Decomposition of the domain

In Figure 4, we have drawn an irregular geometry, which is covered by a
number of rectangles.

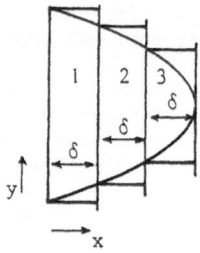

Figure 4 Irregular geometry covered with rectangles.

These rectangles have been chosen such that they all have equal width δ. This allows us to put them together into one large rectangle as drawn in Figure 5.

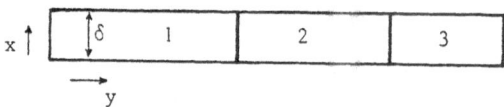

Figure 5 Composition of the rectangles into
one large rectangle.

On this rectangle again a fast calculation can be performed. It will be clear that the rectangles in Figure 4 should have a small overlap in order to calculate differences near the interfaces. This involves some copying of data each time step.

Two numberings

Another possibility is to use two numberings, one in x-direction and one in y-direction as drawn in Figure 6.

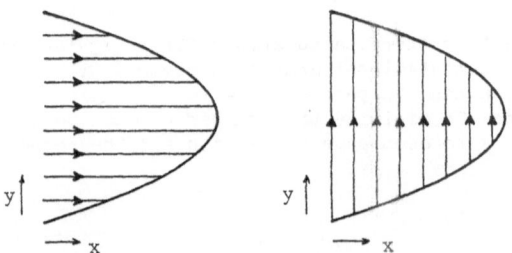

Figure 6 A numbering in x- and y-direction.

Now, $\vec{F}$ is split up into two parts

$$\vec{F} = \vec{F}^x + \vec{F}^y, \qquad\qquad (28)$$

where $\vec{F}^x$ and $\vec{F}^y$ contain only differences with respect to x and y, respectively. During the F evaluation, the numbering of the variables has to be changed in order to have a fast evaluation of both $\vec{F}^x$ and $\vec{F}^y$. The renumbering can be done with a gather or scatter vector instruction.

CONCLUSIONS

In this contribution, we have considered the performance of an explicit method and an implicit method for the SWE's. Implementing an explicit method on a vector computer is quite straight-forward, whereas with implicit methods, one has to reconsider the algebraic solution process in order to solve the occurring set of equations as effective as possible. The stabilization technique presented in this contribution works well for the SWE's, i.e., the explicit Runge-Kutta method with stabilization is competitive with the ADI method with respect to both accuracy and computational costs. Furthermore, it can be easily added to an existing explicit time integration process based on the method of lines approach.

REFERENCES

[1]   Houwen, P.J. van der, and J.G. Verwer, One-Step Splitting Methods for Semi-Discrete Parabolic Equations, Computing 22, pp 291-309, 1979.

[2]   Jameson, A., and D. Mavriplis, Finite Volume Solution of the Two-Dimensional Euler Equations on a Regular Triangular Mesh, AIAA 23rd Aerospace Sciences Meeting, AIAA-85-0435, Nevada, 1985.

[3]   Lambert, J.D., Computational Methods in Ordinary Differential Equations, Wiley, London-New York, 1973.

[4]   Stelling, G.S., On the Construction of Computational Methods for Shallow-Water Flow Problems, Thesis, TH Delft, 1983.

[5]   Wubs, F.W., Stabilization of Explicit Methods for Hyperbolic Initial-Value Problems, in preparation, C.W.I., Amsterdam 1985.

[6]   Wubs, F.W., Smoothing Techniques for Initial-Boundary-Value Problems, in preparation.

METHODS FOR OPTIMISATION AND ACCELERATION OF AN EXPLICIT NAVIER-STOKES CODE WITH APPLICATION TO SHOCK/BOUNDARY-LAYER INTERACTION

E. Katzer
DFVLR-AVA, Institut für Aeroelastik,
Bunsenstraße 10, D-3400 Göttingen, Germany

and

M. Dowling
Institut für Angewandte Mathematik der
Technischen Universität Braunschweig,
Pockelstraße 14, D-3300 Braunschweig, Germany

SUMMARY

The two parts of this paper discuss techniques for improving the programming efficiency of explicit algorithms on the CRAY-1 vector computer at the FORTRAN and ASSEMBLER levels respectively. The high degree of modularity was essential for both simplifying the application of the techniques of vectorisation and for calculating the maximal execution speed of the algorithm and achieving this speed using assembly language.

At the FORTRAN level increasing vector lengths by reducing the dimension of arrays achieved a performance improvement of between 10 % and 50 %, the module coded in assembly language achieved a further improvement of 37 %. The FORTRAN coded program attained an average speed of 60 MFLOPS.

INTRODUCTION

This paper is concerned with techniques for improving the efficiency of an explicit algorithm for solving the Navier-Stokes equations. A high degree of modularity was found to be vital for applying vectorisation techniques, calculating and achieving maximal execution speed using ASSEMBLER, and attaining portability. The capability of the numerical algorithm is demonstrated by its application to shock/boundary-layer interactions and the physical results are shown.

This paper comprises two independent parts; the first part by E. Katzer is concerned with code development at the FORTRAN level and the second part by M. Dowling deals with improvements on the ASSEMBLER level. All numerical calculations were performed on the CRAY-1S of DFVLR in Oberpfaffenhofen. Most of the calculations used the FORTRAN compiler CFT 1.13 under the operating system COS 1.13. ASSEMBLER calculations were carried out with the CAL 1.13 ASSEMBLER.

Physical Problem

To illustrate the capabilities of the algorithm, we present some
numerical results [1] on shock/boundary-layer interaction. The
interaction of a shock wave with a boundary layer is an impor-
tent aspect of aerodynamic flows, as it impairs the performance
of transonic aircraft, turbomachines and reentry vehicles.

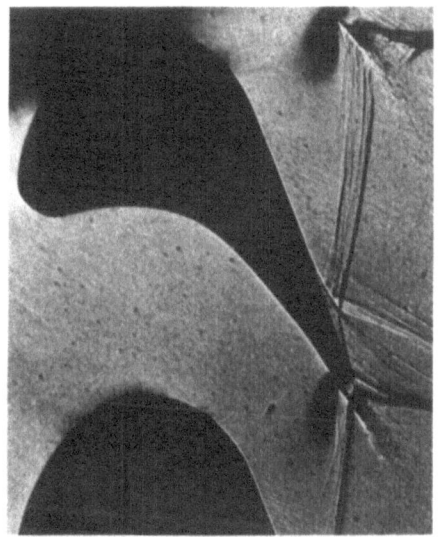
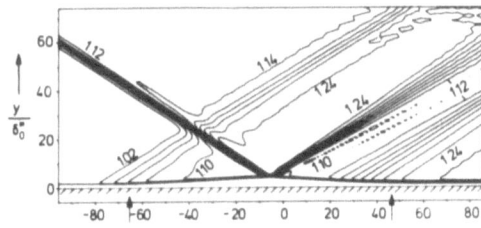
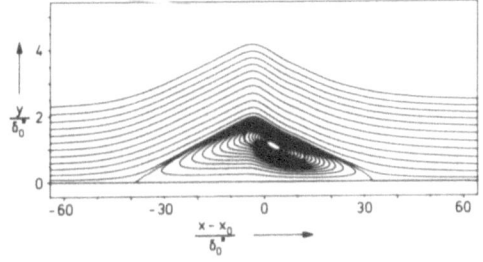

Figure 1a: Interaction of a
shock wave with a laminar
boundary layer in a tran-
sonic turbine cascade;
Experiment [2]

Figure 1b: Laminar shock/boundary-
layer interaction on a flat plate.
Lines of constant density and
streamlines within the boundary
layer; Navier-Stokes solution [1]

Figure 1a shows the interaction of an oblique shock with a laminar boundary layer in turbomachine cascades [2]. The interaction
on a flat plate, which was used as a model problem [1], is shown
in Figure 1b. The isochors in the upper part show the reflection
of the oblique shock as a complicated structure of compression,
expansion and compression zones. The streamlines in the lower
diagram show the boundary layer and an asymmetric structure
within the separation bubble.

Extensive numerical studies have led to a similarity law for the
length of the separation bubble [1]. This enables quick estima-
tion of the extent of the interaction region and can be utilised
to generate appropriate starting values for subsequent numerical
calculations.

Basic Equations and Numerical Algorithm

The governing equations for two-dimensional compressible fluids are the Navier-Stokes equations. For a fixed volume element Vol with surface ∂Vol, the conservation of mass, momentum and energy are:

$$\frac{\partial}{\partial t} \int_{Vol} \rho \, dvol = - \int_{\partial Vol} \rho (\vec{v} \cdot ds) \ , \tag{1}$$

$$\frac{\partial}{\partial t} \int_{Vol} \rho \vec{v} \, dvol = - \int_{\partial Vol} \rho \vec{v} (\vec{v} \cdot ds) + \int_{\partial Vol} S \, ds \ , \tag{2}$$

$$\frac{\partial}{\partial t} \int_{Vol} \rho \, e_t \, dvol = - \int_{\partial Vol} \rho \, e_t (\vec{v} \cdot ds) + \int_{\partial Vol} (S\vec{v} - \vec{q}) \cdot ds \ , \tag{3}$$

with density ρ, velocity $\vec{v}$, total energy e_t, Newtonian stress tensor S and Fourier's low of heat flux $\vec{q}$.

A time-stepping method is used to solve these equations. Initial values, boundary conditions and the necessary grid resolution are discussed in [1]. The numerical method is a finite-volume version of the explicit time-split scheme of MacCormack [3]. For example, a discrete approximation of the continuity equation is given by

$$\rho^{t+\Delta t} = \rho^t - \frac{\Delta t}{Vol} \sum_{\partial Vol} \rho \, \vec{v}^t \cdot \Delta s \ , \tag{4}$$

where sigma (Σ) denotes a discrete approximation of the surface integral. This equation is applied to the predictor and corrector step of the MacCormack scheme, see [1] for details.

From Eq.(4) it is clear that flow variables at the new time level $t + \Delta t$ are dependent only on the variables at the former time level t. They are independent of their neighbouring elements at time level $t + \Delta t$. Therefore, calculations at the individual grid points can be made simultaneously and the algorithm can be entirely vectorised.

Structured Programming and Vectorisation

To increase the efficiency of code development and debugging,
the principles of structured programming have been strictly followed. Although the FORTRAN language is not well suited to support structured programming, some rules have been applied which
are listed here:

- The numerical algorithm is split into several tasks and subtasks, each of which is performed in separate subroutines.

- These subroutines are not referenced within DO-loops; subroutines contain large loops over the entire grid.

- Within these loops, identical tasks are performed at each
  grid point. One is not tempted to perform several tasks
  within one loop and then switch between them using IF-
  statements, as this would inhibit vectorisation.

- Hardware-dependent statements are concealed in separate subroutines. There is a different version of these routines for
  each machine.

- All variables are passed to or retrieved from subroutines
  via the parameter list. COMMON is not used. This enables
  the same subroutine to be used for the predictor and corrector step of the MacCormack scheme by passing different
  starting array elements as actual parameters to dummy arguments. In the subroutine these dummy arguments are de-
  clared to be arrays.

Consequently, the Navier-Stokes program comprises about 50 subroutines. The main part of the algorithm includes 10 subroutines
in which 87% of the numerical work is done. (The amount of work
has been estimated using the flow-trace option of the CFT compiler with inhibited vectorisation (OFF = V)). Each module is
very short, typically 10 to 20 executable FORTRAN statements
representing 10 to 50 floating-point operations per grid point.

A disadvantage of this high modularisation is the overhead introduced by calling a subroutine. To estimate this overhead, computing times for calling modules consisting of a subroutine head
immediately followed by a RETURN statement have been determined.
The calling times depending on the length of the parameter list
are given in Table 1. In practical calculations with grid sizes
of about 10,000, 10 to 50,000 floating-point operations are performed per subroutine call and the calling overhead can be neglected.

The main advantage of the high degree of modularisation is the
increased comprehensibility of the program and the reduced complexity of the subroutines. This simplifies vectorisation, which
in most cases could be done by the autovectorising capabilities
of the compiler, and resulted in an efficient code. Even manual
vectorisation was a quick and easy task, in which

- the formated output was eliminated,
- a FORTRAN statement function was replaced by an in-line code,
- an IF-statement within a loop was exchanged with an CRAY CVMGP function.

Furthermore, its simple structure enables ASSEMBLER coding of some modules, which will be discussed in the second part of this paper.

The good autovectorisation capabilities of the code ensures port-ability as well as compatability with scalar machines. Therefore, testing and calculation of minor problems can be performed on standard computers if a vector computer is not available.

One-Dimensional Array Processing

As we have seen, vectorisation of an explicit algorithm is straightforward. A further increase in computational speed can be attained by increasing the vector length. Clearly this accel-eration depends on machine characteristics, e.g. Hockney's $n_{1/2}$, and will be greater on memory-oriented machines such as CYBER 205 than on vector-register machines.

Since the CRAY CFT compiler vectorises only the innermost DO-loops, the vector length could be increased drastically by changing two-dimensional arrays into large one-dimensional ar-rays. Two-dimensional arrays are stored columnwise in contigu-ous elements of memory and therefore a double DO-loop:

```
          DIMENSION X(N,M), Y(N,M) Z(N,M)
          DO 99    I = 1,N
          DO 99    J = 1,M
             Z(I,J) = X(I,J) * Y(I,J)
   99     CONTINUE
```

can be changed into:

```
          DIMENSION X(N*M), Y(N*M), Z(N*M)
          DO 99    L = 1, N*M
             Z(L) = X(L) * Y(L)
   99 CONTINUE
```

This technique increases the vector length from M to N*M and thus the computational speed. Furthermore, bank conflicts are avoided. Table 2 shows the increase of computational speed for a sample problem of multiplication of two arrays of dimension (47 x 64). In a two-dimensional array-processing calculation, speed depends on rowwise or columnwise calculation and on array dimensioning. For a fairly large vector length of 47, two-dimen-sional array processing is 50% slower than one-dimensional. In the case of a column length of 64, bank conflict occurs which decreases speed further. Even for a vector length of 64, which corresponds to the length of the vector registers of the machine, two-dimensional processing is still 33% slower than one-dimen-sional. Even in scalar mode of the compiler (OFF = V) the two-dimensional processing remains 12 to 37% slower than one-dimensional.

It should be mentioned that, even on a scalar machine (SIEMENS 7865), rowwise processing is 24% and columnwise processing 2% slower than one-dimensional calculations.

In numerical flow simulations, time-stepping operations (Eq.4) are performed at inner points of the flowfield. To apply one-dimensional array calculations, care has to be taken at boundary elements. Here, the numerical calculation leads to physically nonsensical values. In a subsequent step these values are replaced using the boundary conditions. This results in additional numerical work at boundary points, but because the number of boundary points is small compared to the inner points, this overhead can be neglected. To estimate the gain in computational speed in practice, a comparision between one- and two-dimensional array processing was performed using six modules of the Navier-Stokes program, see Table 3. This represents 60% of the numerical work. CPU times were estimated using the flow-trace option of the CFT compiler (ON = F). Depending on the grid dimensions, one-dimensional calculations were about 10% to 20% faster than two-dimensional calculation. Note that all grids consist of approximately the same number of grid points.

Furthermore, one-dimensional processing of multi-dimensional arrays permits a more flexible grid-point distribution. Only the total number of grid points is limited by the DIMENSION statement and not their distribution in the various dimensions. The extension to three-dimensions is straightforward. Even some of the subroutines do not depend on the physical dimensions of the flow. Therefore a one-dimensional array-processing procedure reduces coding and debugging costs by upgrading a two-dimensional flow-calculation program to three dimensions.

Results

As a result of the principles

- high modularisation and
- one-dimensional array processing,

high computational speed on vector as well as scalar computers has been achieved. Table 4 shows CPU times on various machines for 500 time steps and a small 41 x 41 grid. The CRAY FORTRAN compiler CFT 1.10 with operation system COS 1.11 was used. An increase in speed of about a factor of 30 relative to IBM has been realized on the CRAY. In practical calculations on large 151 x 101 grids, the CPU time per time step and grid point (RDP) is 12 microseconds (using CFT 1.13 and COS 1.13). This yields an extremely high computational speed of 60 million floating-point operations per second, i.e. almost one operation per processor cycle of $12.5 \cdot 10^{-9}$ sec. Furthermore, portability to scalar machines has been retained.

In the subsequent part, improvements of the calculation speed by using ASSEMBLER coding are discussed.

Table 1: CALL overhead; CPU time for subroutine call and
        immediate return

length of parameter list	5	9	16	32
$\dfrac{\text{CPU}}{\text{CALL}}$ in 10^{-6} sec	1.5	1.9	2.7	3.2
$\dfrac{\text{cycles}}{\text{CALL}}$	120	150	220	260

Table 2: CPU times for one- and two-dimensional array
        processing. Arrows indicate rowwise (→) or
        columnwise (↓) multiplication; $Z_{ij} = X_{ij} * Y_{ij}$

CPU time in 10^{-3} sec, 1-D ratio	47 $\boxed{\overset{64}{\longrightarrow}}$	47 $\boxed{\overset{64}{\downarrow\downarrow\downarrow}}$	64 $\boxed{\overset{47}{\rightrightarrows}}$	64 $\boxed{\overset{47}{\downarrow}}$	$\overset{3008}{\boxed{\longrightarrow}}$ 1-D
Vector length	64	47	47	64	3008
Vector mode ON = V	0.17 1.35	0.19 1.50	0.57 bank conflict	0.17 1.33	0.13 1.00
Scalar mode OFF = V	1.28 1.36	1.08 1.14	1.30 1.37	1.06 1.12	0.94 1.00

Table 3: Increase of speed of one-dimensional array processing
in practice.
CPU time for six Navier-Stokes modules.

CPU 4000 calls	Gridsize		
	41x41	51x33	63x27
2-D	14.5	13.6	13.0
1-D	11.8	11.8	11.7
2-D/1-D	1.22	1.16	1.11

Table 4: Computation times of Navier-Stokes code on scalar- and
vector computers.
Note that this is a sample calculation with only 500
time steps on a small 41 x 41 grid, which is not sufficient for practical calculations.

	SIEMENS 7865	IBM 3081-D	CRAY-1S vec. inhibited	CRAY-1S vectorised
CPU [sec]	1745	363	70.7	12.5
RDP [10^{-6}sec]	2000	430	84	15
performance	140	29	5.6	1

MAXIMAL PERFORMANCE ON THE CRAY-1 USING CAL

Summary

For a given algorithm on a given computer there is a code for which the CPU time is minimal. For most computers, however, it is neither obvious what this maximal performance is, nor is it obvious how to achieve it. In this short note, we indicate how this can be done for certain, simple, completely vectorised algorithms on the CRAY-1 computer using the CRAY assembly language.

Software Performance

In their now classical work [6], Hockney and Jesshope introduced the pair of numbers r_∞ and $n_{1/2}$ describing the performance of vector computers. These numbers are defined by linearising the performance curve:

$$t = r_\infty^{-1}(n + n_{1/2}) \quad , \tag{5}$$

where t is the time taken to perform an operation on vectors of length n, which must be large. r_∞ is measured in MFLOPS, while $n_{1/2}$, the amount of arithmetic that might have been performed in the time taken to attain half the maximal performance, is measured in FLOP. Hockney and Jesshope compute r_∞ and $n_{1/2}$ for the CRAY computation at primitives; the unary operation of reciprocal; the binary operations of addition and multiplication, and for the ternary operations: uv+w, cu+v, uv+c, where u, v, w are (long) vectors and c is a scalar. In this manner Hockney and Jesshope could make a useful description of the CRAY-1 vector functional units, while at the same time retaining simplicity.

It is possible, however, to define r_∞ and $n_{1/2}$ analogously for any completely vectorised program module, where all the vectors have a single length, n. (We consider only these modules in which all vectors have the same vector length.)

If we deem program modules as being equivalent if, given the same input, they invariable produce the same output (stability questions being ignored), then the greatest, lower bound on r_∞^{-1} for the differing codes within the same equivalence class defines the theoretical, asymptotical maximal performance for the algorithm, denoted by R_∞. R_∞ is then a property of both the algorithm and the computer on which it is executed, but is independent of the software used in coding the algorithm.

If we refine the above equivalence class by further specifying the algorithm be coded in FORTRAN, one defines new value, R_∞^{CFT}, which is the asymptotical maximal performance obtainable for the algorithm when coded in FORTRAN. The ratio R_∞/R_∞^{CFT} is then a measure of the efficiency of the CRAY FORTRAN compiler for the given algorithm.

146

If R_∞ (resp. R_∞^{CFT}) are computable for a given algorithm, one obtains an objective measure for the software efficiency of a given code (resp. FORTRAN code). The objective here is to show how to compute the former for simple, completely vectorised program modules and to give ASSEMBLER programming principles used in attaining this value. It is however, usually difficult to determine R_∞^{CFT} convincingly, although for certain simple codes it is difficult to imagine that it has not been obtained.

Remark:
One might equally well define $N_{1/2}$ as being the minimum value of $n_{1/2}$ achieved by codes within the above equivalence, and ask as to its value and as to whether both R_∞ and $N_{1/2}$ can be obtained for the same algorithm. In fact, the value of $N_{1/2}$ is largely outside the control of even the ASSEMBLER programmer. For coded versions of a given algorithm its $n_{1/2}$ is determined by the $n_{1/2}$ of the functional units used, and, more importantly, the macros used in transfering the arguments to the subprogram, which in turn, depend on the conventions used by the FORTRAN compiler. (These questions have therefore been ignored; however suffice it to say that for TEUNLS described below, it is of the order of about one hundred FLOP. !!!)

Computing R_∞ for Simple Programs

In order to estimate R_∞, a few facts concerning the CRAY-1 architecture must be considered.

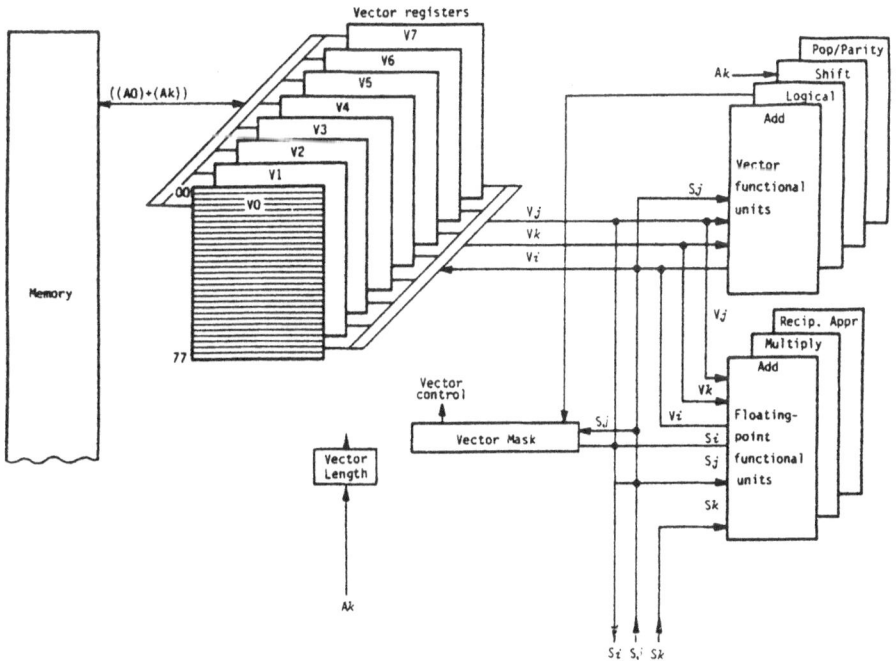

Figure 2: CRAY-1 architecture [9]

147

The vector section of the CRAY-1 computer consists of:

i)   8 vector registers, each containing 64 64-bit words
ii)   a single, segmented memory access port
iii) three floating point, segmented functional units
iv)   four segmented, fixed point functional units.

Functional units take their operands from the vector registers,
and deliver their results to a vector register at a rate of one
value per machine cycle (i.e. 80 MFLOPS)! The vector registers
are loaded from and stored to memory in seperate operations.
For all programming intents and purposes, the memory access
port should be considered as being segmented functional unit.

Each functional unit is capable of operating independently of,
and hence concurrently with each other. This facility is limited solely by data flow dependencies; operand and result registers being reserved for the duration of the execution of a vector instruction, thereby requiring an instruction, requiring a
result from a previous instruction as an operand, to wait until
that previous instruction has completed its execution. Chaining
is the only exception to this rule.

Remark:
The CRAY X-MP is equipped with three memory ports per cpu; two
for loading vector registers from memory, and one for storing
vector registers to memory (among other functions as well).

Chaining

Each vector register has a hardware pointer which points to the
next operand to enter the functional unit in the case of operands, or to the next result to arrive from the functional unit
in the case of result registers. Where a vector register is to
be used both as an operand and as a result in different instructions, the pointer positions for both functions must
agree. I follows that there is exactly one clock period during
which a result register can be used as an operand in starting
a subsequent instruction; otherwise the reservation of the result register takes effect and the subsequent instruction is
forced to wait for the completion of the previous instruction.
When a vector register is used both as a result and as an operand register, the respective functional units are said to
chain.

Clearly, the FORTRAN programmer has no direct control over the
timing of the issue of individual, machine level instructions;
therefore, in FORTRAN, chaining is an haphazard affair.

Remark:
A fundamental innovation of the CRAY X-MP is that now vector
registers are equipped with two element pointers; one for each
of the functions above. The only restriction is that it is for-
bidden that the read pointer precede the write pointer. Chaining is thus no longer an issue on the X-MP.

Memory Access

The basic programming strategy for any computer in which the various functional units are capable of concurrent execution is to identify that functional unit most in demand, to optimise its execution, and to attempt to perform all other operations simultaneously. This strategy is not always possible, but for the CRAY-1 it seldom fails. This is a consequence of its single, most important design defect: it has only one memory access port. A single binary operation requires three vector registers: two operand registers and one result register, so that, even if some intermediate results can be retained in the vector registers, thereby saving a load and a store operation, memory access operations nevertheless almost invariably dominate the computation, and so usually for the sole limiting factor for the performance. Although exceptions to this rule are easy to invent, they seldom appear in practice. The programming strategy is then obvious: utilise the memory access port continuously; avoiding if at all possible, the necessity to store and reload intermediate results, and perform all other operations simultaneously with the memory access port. If one can achieve this, it is clear that the theoretical best possible performance, R_∞, has been obtained.

Remarks:
1. Since there are only eighth vector registers, the temporary storing of intermediate results to memory is sometimes inescapable. This complicates the strategy considerable as the question then arises as to which register is to be temporarily stored to minimise the execution time. The possibilities are, however, fortunately limited.
2. The above mentioned architectural defect for the CRAY-1 has been obviated on the X-MP by the introduction of two additional memory access ports; two ports (among their other duties) are for loading vector registers from memory; and the third (among its other duties) stores vector registers to memory. (These other duties cannot be ignored by the assembly language programmer; for example: an instruction buffer re-load at the wrong moment can have an adverse effect on the execution effiency of the code).
3. That the functional unit most in demand be memory access is not crucial, and for the the CRAY X-MP it is not generally the case. Vital is only that this functional unit be identified and optimised.
4. An unpleasant consequence of this strategy is a severe loss of comprehensibility; memory access operations are scattered throughout the code, occuring at times required for optimisation, but not in the sequential order that might logically be expected. Documentation using comments is therefore of paramount importance.
5. To ameliorate the problem of having insufficient vector registers for intermediate results, it is often desireable to suppress chaining deliberately. This is because chaining requires the use of five out of the eight vector registers where as a non-chained, binary operation requires only three.

The Computation of R_∞

The method of computation will be illustrated by means of a simple example, namely performing the ternary operations:

$$z = c \cdot x + y \quad ,$$

where $x, y, z \in \mathbb{R}^n$, $c \in \mathbb{R}$.

Step 1 Identify the functional unit most in demand:
Here, there are three vector memory access operations:
i)   load x into a V-register
ii) load y into a V-register
iii) store z into memory.
There is also one floating point addition, and one floating point multiplication, which could possibly be chained. Memory access is, then, the most used functional unit.

Step 2 The naive calculation:
Since there are sixty four elements in a vector register, one works in sixty four element sub-vectors. For such sub-vectors, there are 2 x 64 = 128 floating point operations. There are three memory access operations that must be performed sequentially. They require therefore 3 x 64 = 192 machine cycles. Since each machine cycle is 12.5 nsec, the estimate for R_∞ is

(128 x 80) / 192 = 53 1/3 MFLOP   .

Step 3 The exact estimate:
The above estimate assumes (that during the machine cycle following that in which the last element of the previous memory access operation has set off along its segmented path to memory,) the first element of the following memory access instruction sets off along the same path.
The fact is, however, that the entire path must be empty before the next memory access instruction can issue; since this path contains six segments, the next instruction must wait a further six clock periods, so reducing the above estimate for R_∞ to

(128 x 80) / (3 x (64+6)) = 48.76 MFLOPS   .

Approaching the Estimate Value of R_∞ using CAL

The first phase of CRAY assembly language programming is to code the vector operations naively, and then supplement these with the necessary scalar adressing operations which should be performable concurrently with the previous vector instruction.

The second phase consists of running CRAY program CYCLES on this code to check in instruction issue/instruction issue delay timings for the individual instructions. The code is then modified accordingly. Phase two is repeated as often as necessary to obtain an optimal code.

For the above example, a simplified first phase code might be the following:

```
        A.N.    n                   Address register A.N. contains the
                                    vector length n
        VL      A.N                 Set the vector length register
LOOP    =       *                   LOOP is the instruction address of
                                    loop starting position
        S.C.    c                   Load scalar register S.C with the
                                    scalar
        V.X     x                   Load vector register V.X with x
        V.PROD  S.C*FV.X            Scalar mult. c·x stored in V.PROD
        V.Y     y                   Load V.Y with the vector y
        V.Z     V.PROD+FV.Y         c·x+y in V.Z (vector addition)
        z       V.Z                 Store V.Z to memory
        A.N     A.N-0'100           Reduce Vector length by $64_{10}=100_8$
        VL      A.N
        AO      -A.N
        JAM     LOOP                Go to loop if there is more work
                                    to do
```

Running CYCLES on this code shows that the instruction to load V.Y and the following floating point addition operation cannot chain; eight cycles after the issue of the load instruction for V.Y. the chain slot time for memory access occurs; at this stage, however, the last element of V.PROD still remains to be computed so that the reservation on V.PROD remains in force missing the chain slot time by one clock period and thus delaying the entire computation seventy two clock periods.

An improved version of this code will force the instruction issue for loading V.Y to be delayed by a further cycle to allow chaining with the subsequent vector instruction. (Its instruction issue time is already delayed by 56 CPs owing to the current execution of the load instruction for V.X.)

A second failing with this code is that the memory access port is idle for eight clock periods (the chain slot time for floating point addition) between loading the last element of V.Y from memory and the instruction issue for storing V.Z to memory (owing to the reservation on V.Z)! This latter failing is overcome by starting a second, identical loop immediately by reloading V.X with the next sixty element segment and storing V.Z to memory at a later stage; the size of the code lenght being doubled. This is typical of the phased computation of CRAY assembly language programming in which different phases are being performed concurrently.

Subroutine TEULNS

As a practical application of these principles, it was decided to code the following FORTRAN subroutine in assembly language. This subroutine is one of the sub-program modules belonging to my colleague's explicit Navier-Stokes code.

```
      SUBROUTINE TEULNS(NCEL,
     &                        U,V,P,RO,RU,RV,RE,
     &                        EOX,EUX,EVX,EEX,
     &                        EOY,EUY,EVY,EEY)
      DO 357 l=1,NCEL
      EOX(L)=-RO(L)*U(L)
      EUX(L)=EUX(L)-RU(L)*U(L)+P(L))
      EVX(L)=EVX(L)-RV(L)*U(L)
      EEX(L)=EEX(L)-(RE(L)+P(L))*U(L)
      EEY(L)=EEY(L)-(RE(L)+P(L))*V(L)
      EUY(L)=EUY(L)-RU(L)*V(L)
      EVY(L)=EVY(L)-(RV(L)*V(L)+P(L))
      EOY(L)=-RO(L)*V(L)
  357 CONTINUE
      RETURN
      END
```

Inspections shows that there are

- 21 memory access operations
- 11 floating point additions or subtractions
- 8 floating point multiplications

hence
- 19 floating point operations

performed during each execution of the above loop.

Once again, memory access dominates and so is the first concern. In fact it was possible to code the above subroutine in assembly language without having to resort in storing intermediate results temporarily to memory. During the entire execution, the memory access port was in continuous operation and thus was the sole consideration determining the execution speed.

Results

The results can be summarised:

The theoretical value of R_∞ is 68.12 MFLOPS.
The FORTRAN version of the code achieved R_∞^{CFT} = 49.1 MFLOPS.
The CAL version achieved 67.0 MFLOPS.

Since this subroutine is responsible for only 10 % of the total computations of the Navier-Stokes program the overall acceleration is of the order of 3 %.

Remark:
A similar analysis to that above has been made for an assembly language code for TEULNS on the CRAY X-MP with one processor. Here, the floating point addition functional unit is the sole factor restricting the execution speed of the program. Considerable care must be taken to avoid using registers that are still reserved as a result of a previous vector instruction.

For both the CRAY-1 and the CRAY X-MP the number, eight, of vector registers is a severe restriction for the programmer. The estimated theoretical asymptotical maximum performance of the CRAY X-MP for TEULNS is 166.23 MFLOPS.
It should be noted here that the X-MP code represents an improvement factor of about 2.5 compared with the CRAY-1. This can be compared with FORTRAN codes which appear to improve by a factor of somewhere between 1.4 and 1.8 [4]. This is conjecturally a result of above mentioned point over which ASSEMBLER programmer has control, but the CFT compiler probably not.

REFERENCES

[1]   KATZER, E.: Numerische Untersuchung der laminaren Stoß-Grenzschicht-Wechselwirkung.
DFVLR-FB 85-34 (1985).

[2]   GRAHAM, C.G., KOST, F.H.: Shock Boundary Layer Interaction on High Turning Transonic Turbine Cascades.
ASME Publ. Nr. 79-GT-37 (1979).

[3]   MacCORMACK, R.W., BALDWIN, B.S.: A Numerical Method for Solving the Navier-Stokes Equations with Application to Shock-Boundary Layer Interactions.
AIAA Paper 75-1 (1975).

[4]   DETERT, U.: Performance Comparison for CRAY-1/S and CRAY X-MP by Means of FORTRAN Kernels und User Programs.

[5]   DETERT, U.: Vector Processing on CRAY-1 and CRAY X-MP Workshop on "Use of Supercomputers in Theoretical Science" at University of Antwerp, July 1984.

[6]   HOCKNEY and JESSHOPE: Parallel Computers: Architecture, Programming and Algorithms.
Bristol, 1981.

[7]   HOCKNEY: $(r_\infty, n_{1/2}, s_{1/2})$ Measurements on the 2-CPU CRAY X-MP, to appear in "Parallel Computing", Vol.2, No.1.

[8]   WIESE, G.: Entwicklung, Laufzeitanalyse und Optimierung von CRAY-ASSEMLBER-Programmen.
Diplomarbeit, Inst.Angew.Mathematik, Abt.Rechnentechnik; Technische Universität Braunschweig, 1983.

[9]   CRAY-1 Computer Systems, S Series
Mainframe Reference Manual, HR-0029,
CRAY Research Inc., 1982.

ON VECTORIZATON OF A 2.D. NAVIER-STOKES SOLVER

Y. LECOINTE & J. PIQUET
Computational Fluid Dynamics Group
E.N.S.M. Laboratoire d'Hydrodynamique Navale
1 ,rue de la Noe
F-44072 NANTES Cedex

SUMMARY

This paper is devoted to the implementation of several vectorization techniques for the resolution of 2 D Navier-Stokes problem. The original version of the code was running on classical scalar computers and the main part of this work is the adaptation of the algorithms used for vectorization. In the first part, the equations to be solved are presented; the numerical methods (higher order finite differences)and the algorithms are presented in the second part. The adaptation for vectorization is also detailed. In the third part, the results concerning the speed-up got by the vectorization and by the optimization of the different versions of the codes are presented; endly, some comparisons between a scalar versus a vectorial computer are done.

INTRODUCTION

The code to be presented here is relative to the two-dimensional Navier-Stokes equations written in their vorticity streamfunction formulation and is used to compute unsteady flows around several geometries like circular or elliptic cylinders or airfoils. The original scalar mode used on scalar computers was not fully optimized. All the tests were performed on the NAS-9080 of CIRCE or on the CRAY1-S of the C.C.V.R..

In order to preserve in a sense the portability of the code between several computers, CAL language is not used to optimize the speed of the Cray; moreover, when some special Cray functions or subroutines are called, equivalent routines are written in standard Fortran so as to insure if necessary the portability.It is also possible to test the efficiency of these routines compared to standard Fortran routines.

THE EQUATIONS

The non-dimensional Navier-Stokes equations, in their vorticity streamfunction formulation written in a cartesian frame are written as follows :

$$\Psi_{,xx} + \Psi_{,yy} = \omega \quad , \tag{1}$$

$$\partial\omega/\partial t + (U\omega)_{,x} + (V\omega)_{,y} = (\omega_{,xx} + \omega_{,yy})/RE \quad , \tag{2-a}$$

$$\partial\omega/\partial t + U\omega_{,x} + V\omega_{,y} = (\omega_{,xx} + \omega_{,yy})/RE \quad . \tag{2-b}$$

The equations (2-a) and (2-b) are the conservative or the convective form of the vorticity transport equation; where U and V are the x and y components of the velocity defined by:

$$U = -\Psi_{,y} \quad \text{and} \quad V = \Psi_{,x} .$$

RE is the Reynolds number defined by $RE = U a/\nu$ where U_∞ is the upstream velocity, a is a reference length and ν the viscosity of the fluid.

The cordinate transformation consists of two steps: (i) the physical domain round a two-dimensional body is conformally mapped onto the domain outside the unit circle, (ii) the outside this circle is mapped onto a rectangular domain; this transformation is of "O" type. Therefore, equations (1,2,a,b) become:

$$\Psi_{,\zeta\zeta} + \Psi_{,\eta\eta} = g\omega \quad , \tag{3}$$

$$(\partial\omega/\partial t)/g + (-\Psi_{,\eta}\omega)_{,\zeta} + (\Psi_{,\zeta}\omega)_{,\eta} = (\omega_{,\zeta\zeta} + \omega_{,\eta\eta})/RE \quad , \tag{4-a}$$

$$(\partial\omega/\partial t)/g - \Psi_{,\eta}\omega_{,\zeta} + \Psi_{,\zeta}\omega_{,\eta} = (\omega_{,\zeta\zeta} + \omega_{,\eta\eta})/RE \quad , \tag{4-b}$$

where g is the modulus of the transformation: $g = |dz/dZ|$, $z = x + iy$ and $Z = \zeta + i\eta$ are the affixes of a point in the physical domain, respectively in the computational one.

In some situations, it is possible to refine the mesh near the body using stretching functions in each direction. With this technique the separability of the Laplace operator is preserved. The equations in the computational domain are:

$$(f')^2\Psi_{,XX} + f''\Psi_{,X} + (h')^2\Psi_{,YY} + h''\Psi_{,Y} = g\omega \quad , \tag{5}$$

$$\partial\omega/\partial t + f'g'[(-\Psi_{,Y}\omega_{,X}) + (\Psi_{,X}\omega_{,Y})\ ] = [(f')^2\omega_{,XX} + (g')^2\omega_{,YY} + f''\omega_{,X} + g''\omega_{,Y}]/RE, \tag{6-a}$$

$$\partial\omega/\partial t + f'g'[-\Psi_{,Y}\omega_{,X} + \Psi_{,X}\omega_{,Y}] = [(f')^2\omega_{,XX} + (g')^2\omega_{,YY} + f''\omega_{,X} + g''\omega_{,Y}]/RE \quad , \tag{6-b}$$

where $X = f(\zeta)$ and $Y = h(\eta)$ are analytical functions and the computational domain is (X,Y) with constant mesh-size.

NUMERICAL METHODS

The numerical methods used to solve the equations (3) and (4-a,b) or (5) and (6-a,b) are described in details in [1] and are summarized here to understand the way of vectorizing the algorithms used.

155

POISSON EQUATION

The Poisson equation is solved by means of a classical A.D.I. method
with Wachpress'optimized parameters for 8 or 16 iterations. Boundary conditions are needed to solve this equation; in ζ direction ,on the body,the no-
slip condition is imposed and at the outer boundary, the flow is assumed to
be known ;in η direction,as the mesh transformation defined before is of "0"
type, a periodicity condition is prescribed. The spatial discretisation is
h^4 accurate using Numerov formulae for the equation (4) or a Mehrstellen
technique for the equation (6). The Mehrstellen technique was presented by
Collatz in [2] and by Krause et Al. in [3] and the Numerov formula is:

$$(f''_{i+1}+10 \ f''_i + f''_{i-1} \ )/12 =(f_{i+1} -2 \ f_i +f_{i-1})/h^2 \ , \qquad (7)$$

where i stand for ζ or η direction and the double prime the second order
derivative in ζ or η direction.
In both cases we got at each iteration and for each direction a relation
between three adjacent points like the following:

$$a_i \ \Psi_{i-1} + b_i \ \Psi_i + c_i \ \Psi_{i+1} = d_i \ , \qquad (8)$$

where i stands for ζ or η direction;the unknowns of the problem are Ψ and its
first and second derivatives in each direction. When the convergence is
attained, the first derivatives are computed by means of an hermitian formula like:

$$(f_{i+1} - f_{i-1})/h = (f'_{i+1} +4 \ f'_i +f'_{i-1} \ )/3. \qquad (9)$$

The second derivatives are computed by elimination or using other hermitian formulae.
If Nx and Ny are the number of mesh points in ζ and η direction,for each
iteration, Nx-1 systems like (10-a) and Ny-1 systems like (10-b) are to be
solved at each iteration.The coefficients of the first members of these
systems depend only on the geometrical parameters and on the number of
iteration; thus it is possible to pre-process the resolution of these systems to save computational time. In ζ direction,the systems (10-a) are
tridiagonal and in η direction one more row and one more column appeared
because of the periodicity condition (10-b)

$$MX=D_1 \ , \qquad (10\text{-}a)$$

and

$$NX=D_2 \ . \qquad (10\text{-}b)$$

The matrices M and N are the following:

$M=$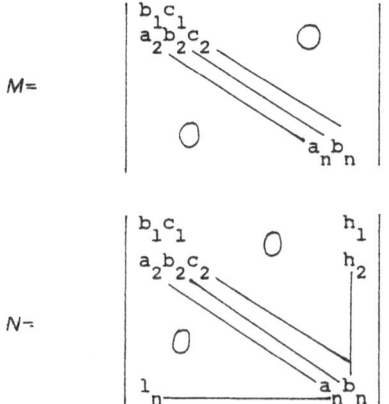

$N=$

Scalar resolution:

Eq.(10-a,b) are solved with the help of a L.U. decomposition for M and N. It can be easily shown that the resulting form for the L and U matrices is:

$M=$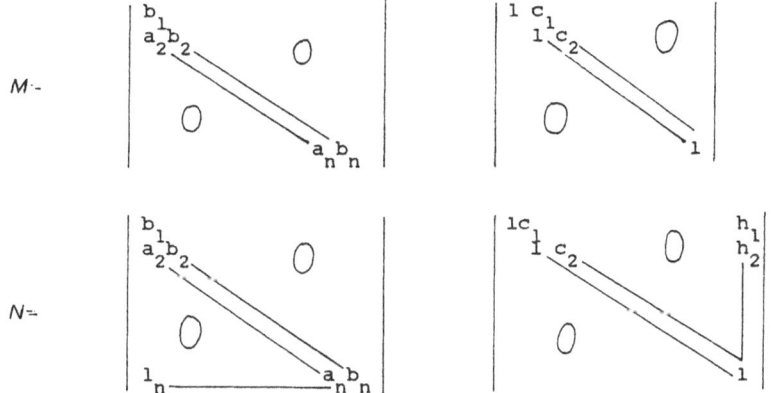

$N=$

The LU factorization is followed by a back substitution; the description of these two phases is given below:
- L.U. factorization:
for the matrix M the coefficients a,b,c are given by:

$$\tilde{b}_1 = b_1$$
$$\tilde{c}_i = c_i / \tilde{b}_i \qquad\qquad i=1,n-1 ,$$
$$\tilde{b}_i = b_i - \tilde{a}_i \tilde{c}_i \qquad\qquad i=2,n ,$$
$$\tilde{a}_i = a_i \qquad\qquad i=2,n ,$$

for the matrix N the coefficient a,b,c,l,h are :

$$\tilde{b}_1 = b_1 \quad , \quad \tilde{l}_1 = l_1 \quad , \quad \tilde{h}_1 = h_1 \quad ,$$

$$\tilde{a}_i = a_i \qquad\qquad\qquad i=1,n$$

$$\tilde{b}_i = b_i - \tilde{a}_i \tilde{c}_i \qquad\qquad i=2,n-1$$

$$\tilde{c}_i = c_i / \tilde{b}_i \qquad\qquad\quad i=2,n-2$$

$$\tilde{l}_i = l_i - \tilde{c}_{i-1}\tilde{l}_{i-1} \qquad\quad i=2,n-1$$

$$\tilde{h}_i = (\tilde{h}_i - \tilde{a}_i \tilde{h}_{i-1})/\tilde{b}_i \qquad i=2,n-1$$

$$\tilde{a}_n = \tilde{l}_{n-1} \quad , \quad \tilde{c}_{n-1} = \tilde{h}_{n-1}$$

$$\tilde{b}_n = b_n - \Sigma(\tilde{h}_i \tilde{l}_i) \qquad (i=1,n-1).$$

-resolution of the systems:
if D define the right-hand side of the systems (10-a,b),the backward substitution is performed in two steps,first LY=D and secondly UX=Y,so for the matrix M,we got:

$$Y_1 = d_1 / \tilde{b}_1$$

$$Y_i = (Y_i - \tilde{a}_i Y_{i-1}/\tilde{b}_i \qquad\qquad i=2,n$$

$$X_n = Y_n$$

$$X_i = Y_i - \tilde{c}_i X_{i+1} \qquad\qquad i=n-1,1$$

and for the matrix N:

$$Y_1 = d_1$$

$$Y_i = d_i - (\tilde{a}_i / \tilde{b}_{i-1}).Y_{i-1} \qquad\qquad i=2,n-1$$

$$Y_n = d_n - \Sigma (\tilde{l}_i / \tilde{b}_i)Y_i \qquad\qquad i=1,n-2$$

$$X_n = Y_n / \tilde{b}_n$$

$$X_i = (Y_i - \tilde{c}_i Y_{i+1} - \tilde{h}_i X_n)/\tilde{b}_i \qquad\qquad i=n-1,1$$

backward-forward dependencies are obviously present : they inhibit the vectorization of the loops.

Remarks:
To compute Ψ in the whole domain at iteration k, one sweep per direction has to be performed and, in each direction, one tridiagonal system has to be solved; the scalar algorithm is :

```
for j=2,Ny-1
for i=2,Nx-1
compute d
          i
next i
solve M Ψ    =d
          i,j  i
next j


for i=2,Nx-1
for j=2,Ny-1
compute d
          i
next j
solve N Ψ    =d
          i,j  j
next i
```

If the loops where the d coefficients are evaluated are vectorizable,the resolution of the systems,as mentionned before,is not vectorizable. The L.U. factorization is neither vectorizable,but as this factorization is made once for all ,it is not time-consuming with respect to the overall CPU time.

Vectorization:

To get an efficient vectorizable code,the algorithm is now written :

```
for j=2,Ny-1
for i=2,Nx-1
compute d
          i,j
next i
next j
solve M Ψ    =d
          i,j  i,j
```

In this case,for the first sweep, the backward substitution is performed with an inner loop in j so that the backward-forward dependencies with respect to i index appear in the outer loop and the inner loop is now vectorizable. The same technique is used for the second sweep.

The main difference between the vectorizable algorithm and the scalar one is the change of the dimension of the local array d; and with only a small increase of the local core needed in this routine,it is possible to speed-up the resolution of these equations. The same technique is used to compute the first derivatives in each direction after convergence.

The results obtained with the resolution of Poisson equation will be presented below.

VORTICITY TRANSPORT EQUATION

Two possibilities are offered to solve the vorticity transport equation;one is the classical A.D.I. method,the second one is based on an approximate factorisation of Beam-Warming type [4]. In connection to A.D.I. method,it is possible to use centered h^2 schemes in space or upwind corrected schemes which is also h^2 ;this scheme was proposed by Ta Phuoc [5].These schemes are refered as Centered Conservative H2 (CCH2) or Upwind Conservative H2 (UCH2). With Beam-Warming factorization,it is possible to use Mehrstellen method which is centered and h^4 accurate or Operator Compact Implicit method which is also h^4 accurate but without theoretical mesh-size limitation; these schemes are refferred to Convective Mehrstellen or Upwind

Operator Compact Implicit (UOCI). This method was propsed by Berger et Al. [6].

For these two methods,we are led to the resolution of two systems(one per direction), like (10-a,b). But here,the coefficients of the two matrices M and N are not only depending on geometrical data but on the flow itself by means of the local velocities;so,in this case,the pre-processing (that is LU factorization) is not made once for all but at each time step.

Scalar code:

In the scalar code,the algorithm is the following:

```
for j=2,Ny-1
for i=2,Nx-1
compute a ,b ,c  and d
          i  i  i      i
next i
solve Mω   = d
       i,j    i
next j
```

and a similar algorithm for the second sweep.
If the computation of the coefficient is vectorizable,here again,the problem of dependencies is present in the resolution of M and N systems;this problem is also present in the L.U. factorization.

Vectorization:

To vectorize this algorithm,we use again the same technique as for Poisson equation but this tecnique is used both for L.U. factorization and backward substitution. The two phases made by two different routines before can be done in the same routine. The algorithm is :

```
for j=2,Ny-1
for i=2,Nx-1
compute a   ,b   ,c    and d
         i,j  i,j  i,j      i,j
next i
next j
solve M ω   = d
         i,j    i,j
```

NUMERICAL RESULTS

The results about vectorization are presented below and are relative to the resolution of Poisson equation and vorticity transport equation. Global results about the code used for the resolution of a classical problem are also given.

A routine called DPOICL or VPOICL is used to solve the Poisson equation, this routine calls other subroutines; the word code is relative to the task of solving this equation. This code was tested for six different meshes and the CPU time spent are given in Table 1 with the speed-up ratio referred to the original code(this code was not optimized). The CPU-times given correspond to the pre-process and the resolution of one equation per mesh. For a real case, the pre-process is made once in the first call of (D)VPOICL.

The original code is not well suited for vectorization, and the vectorization is not very efficient. Some modifications in the order of DO-loops and in the resolution of three-diagonal systems were first implemented in the code "vectorial a". If the speed-up is important in backward substitution, the global ratio for the whole code is lower. In fact, after each half-iteration, the convergence is checked by computing the maximum value of the difference between the iterations. This computation involves a lot of tests and was not vectorized. It is important to notice that some modifications optimize even the scalar version of this code.

If the maximum value is now computed in a separate loop, (code "vectorial b"), the CPU time is decreasing a little; this indicates that tests are time consuming.

In the codes "vectorial c" and "vectorial d", the maximum is computed whith the help of a Fortran subroutine or a CRAY's one respectively. For these two codes, the tests are performed for a 50X100 array which is the upper value of the mesh size, even if the real number of point is lower. For this reason, the results are not significant.

In the codes "vectorial e" and "vectorial f", the maximum value is computed from real computational points; the maximum speed-up ratio is 4.5 with a Fortran subroutines and 8.9 with the Cray's function. This indicates that the functions from library are efficient and that this functions must be used.

A possibility to reduce the number of tests is to check the convergence through the whole computational domain when the convergence is achieved at a chosen point; this possibility should be used in the scalar code. The code "vectorial g" has this implementation. To understand the results obtained in Table 1, they are plotted in figure(1-a) for CPU time and in figure (1-b) for speed-up ratios. The figure (2) gives the ratios of the CPU time of the three vectorized codes with respect to the CPU time without vectorizer. The values are given in Table 2. (The lines drawn between points do not correspond to real values).

Table 1   Resolution of POISSON equation.

Essai \ Maillage	5 x 10		10 x 20		20 x 40		30 x 60		40 x 80		50 x 100	
scalaire	$.1598 \ 10^{-1}$	1	$.5047 \ 10^{-1}$	1	.1526	1	.3121	1	.5289	1	.8033	1
scalaire vectorisé	$.1422 \ 10^{-1}$	1.12	$.4362 \ 10^{-1}$	1.16	.1265	1.2	.2538	1.23	.4251	1.24	.6405	1.25
vectoriel (a) (scalaire)	$.8831 \ 10^{-2}$	1.8	$.3077 \ 10^{-1}$	1.64	.1074	1.42	.2348	1.33	.4130	1.28	.6421	1.25
vectoriel (a)	$.6301 \ 10^{-2}$	2.54	$.1812 \ 10^{-1}$	2.78	$.5452 \ 10^{-1}$	2.80	.1130	2.76	.1946	2.72	.2972	2.70
vectoriel (b)	$.6024 \ 10^{-2}$	2.65	$.1610 \ 10^{-1}$	3.13	$.4518 \ 10^{-1}$	3.58	$.9096 \ 10^{-1}$	3.43	.1554	3.40	.2348	3.42
vectoriel (c)	.1193	.13	.1620	.31	.1686	.90	.1861	1.73	.2077	2.55	.2316	3.47
vectoriel (d)	$.1357 \ 10^{-1}$	1.18	$.2101 \ 10^{-1}$	2.40	$.2745 \ 10^{-1}$	5.56	$.4489 \ 10^{-1}$	6.9	$.6624 \ 10^{-1}$	7.98	$.9048 \ 10^{-1}$	8.88
vectoriel (e) (scalaire)	$.9727 \ 10^{-2}$	1.64	$.2855 \ 10^{-1}$	1.77	$.7983 \ 10^{-1}$	1.91	.2415	1.29	.4241	1.25	.6589	1.22
vectoriel (e)	$.6753 \ 10^{-2}$	2.43	$.1477 \ 10^{-1}$	3.42	$.3355 \ 10^{-1}$	4.55	$.8876 \ 10^{-1}$	3.5	.1511	3.5	.2271	3.54
vectoriel (f) (scalaire)	$.8244 \ 10^{-2}$	1.94	$.2344 \ 10^{-1}$	2.15	$.6421 \ 10^{-1}$	2.38	.1924	1.62	.3368	1.57	.5221	1.54
vectoriel (f)	$.5308 \ 10^{-2}$	3.01	$.9774 \ 10^{-2}$	5.16	$.1807 \ 10^{-1}$	8.44	$.3965 \ 10^{-1}$	7.87	$.6366 \ 10^{-1}$	8.3	$.9025 \ 10^{-1}$	8.9
vectoriel (f) avec test local	$.6692 \ 10^{-2}$	2.40	$.1048 \ 10^{-1}$	4.81	$.1795 \ 10^{-1}$	8.5	$.3637 \ 10^{-1}$	8.58	$.5759 \ 10^{-1}$	9.18	$.8123 \ 10^{-1}$	9.89

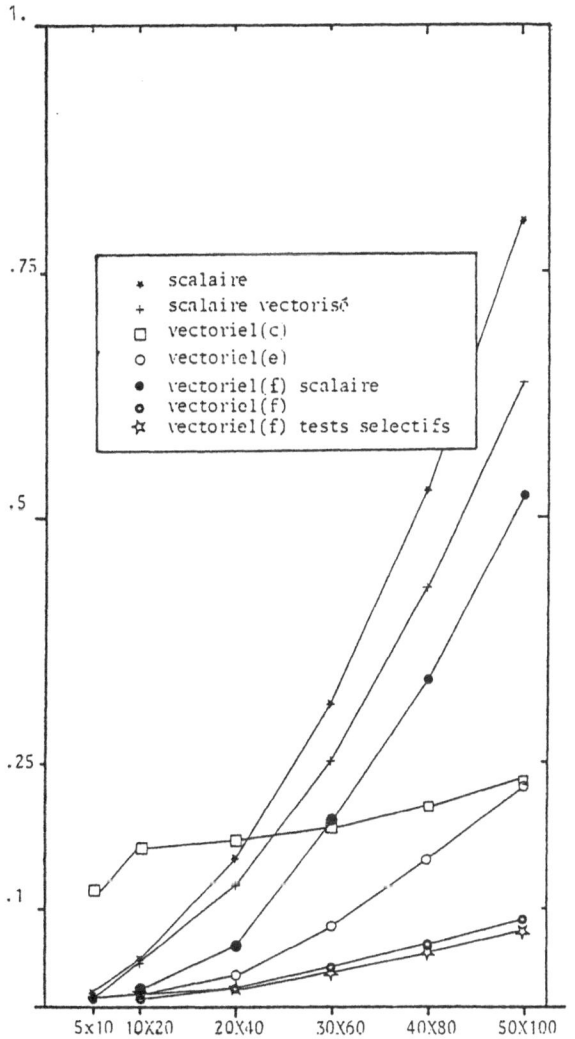

Fig 1-a Resolution of POISSON equation ;
        CPU time

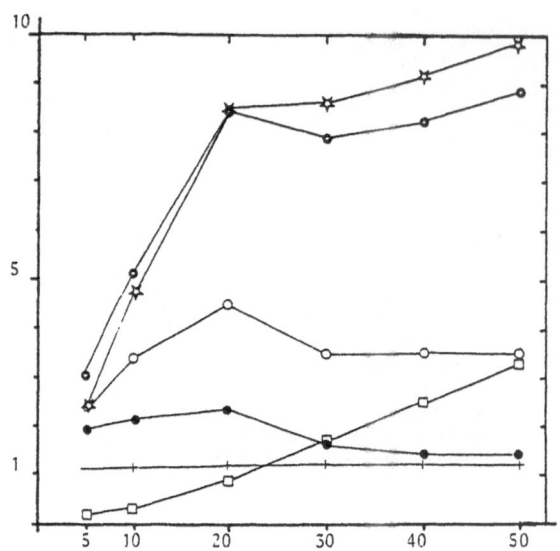

Fig. 1-b POISSON equation Speed-up ratios

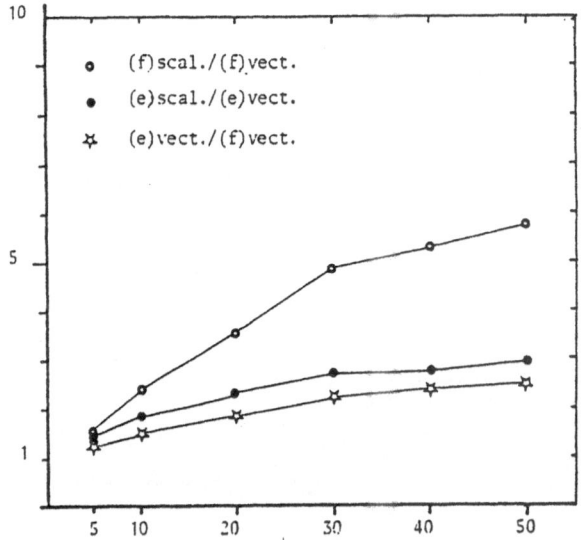

Fig. 2 POISSON equation Comparison between vectorial codes

Table 2 Speed-up ratios for Poisson equation

mesh	5X10	10X20	20X40	30X60	40X80	50X100
scal/scal vect	1.12	1.16	1.20	1.23	1.24	1.25
vect(e)sc/vect(e)	1.44	1.93	2.38	2.72	2.81	2.91
vect(f)sc/vect(f)	1.55	2.40	3.55	4.85	5.29	5.79
vect(e)/vect(f)	1.27	1.51	1.86	2.24	2.37	2.50

The Table 3 presents the Flow-trace for a job which solves a Poisson equation for a 50X100 mesh. For preprocess,DADIOP,DTRI,DLU3 and DLU3L are called,- while for the resolution itself,VRLU3L and VRLU3 are the routines where the speed-up ratios are the more important(9.8 and 9.35 respectively). The comparison between the time used by the code "vectorial f" with and without the vectorizer gives the values of 5.56 and 1.35 .

VORTICITY TRANSPORT EQUATION

The vorticity equation is solved like the Poisson equation by means of some subroutines. The problems here are first the computation of the coefficients of matrices we have to solve and secondly the resolution of the systems. The results presented in Table 4 give the CPU time consumed by two codes "vectorial a" and "vectorial b" with A.D.I. factorization connected to two schemes in space,CCH2 and UCH2.With UCH2 schemes,the main difference between the two codes is the location of the tests;if CCH2 schemes are chosen,the loop where tests are performed is not done. In code "vectorial b" all the tests are performed in a separate loop and the computation of the coefficients a_i,b_i,c_i and d_i is made in an other loop.

The figures (3 a) and (3-b) corresponding to the Table 4 give the plot of CPU-time as a function of the mesh for CCH2 and UCH2 schemes. The Table 5 presents the speed-up ratios plotted in Figure (3-c). The upwinding is roughly 10% more expensive than centered schemes,but the speed-up is bigger (6. instead of 4.). As mentionned before,the vectorizaton of the backward substitution is possible but we have to increase the core requirement;we need local arrays like a(N,N) where N is the maximum dimension of the mesh.

Core requirement:

It is possible to use the capacity of the vector registers of a Cray and to reduce the core requirement using now a(64,n) array;in this case the tests concerning the length of a loop are done with Fortran instructions;- some experiments show us that this reduction of core requirement must be paid by an increasing of the CPU time of roughly 20%. (This value is in fact a function of the length of the loop and of the number of vectorized loops

Table 3 POISSON equation ;Flow trace.

FLOW TRACE --- SUMMARY				FLOW TRACE --- SUMMARY			
ROUTINE	TIME	%	CALLED	ROUTINE	TIME	%	G
1 SPEAR	0.047300	5.85	1	1 SPEAR	0.047199	7.43	1.
2 DCOPIV	0.001786	0.22	77	2 DCOPIV	0.000501	0.08	3.56
3 DPOICL	0.428082	52.98	2	3 DPOICL	0.256928	40.44	1.67
4 DADIOP	0.001586	0.20	1	4 DADIOP	0.001567	0.25	1.
5 DTRI	0.000274	0.03	2	5 DTRI	0.000271	0.04	1.
6 DLU3LC	0.003653	0.48	17	6 DLU3LC	0.003848	0.61	1.
7 DLU3	0.000834	0.10	17	7 DLU3	0.000834	0.13	1.
8 DRLU3	0.125538	15.54	1800	8 DRLU3	0.125515	19.76	1.
9 DRLU3L	0.198744	24.60	882	9 DRLU3L	0.198634	31.27	1.
*** TOTAL	0.807998			*** TOTAL	0.635297		1.27
*** OVERHEAD	0.083737			*** OVERHEAD	0.083663		

original scalar code   vectorized scalar code

FLOW TRACE --- SUMMARY					FLOW TRACE --- SUMMARY			
ROUTINE	TIME	%	CALLED	G	ROUTINE	TIME	%	G
1 VPEAR	0.048318	8.08	1	1.	1 VPEAR	0.048089	33.12	1.
2 DCOPIV	0.001890	0.32	82	1.	2 DCOPIV	0.000533	0.37	3.35
3 VPOICL	0.322302	53.88	2	1.33	3 VPOICL	0.056053	38.60	7.64
4 DADIOP	0.001546	0.26	1	1.	4 DADIOP	0.001527	1.05	1.
5 DTRI	0.000271	0.05	2	1.	5 DTRI	0.000271	0.19	1.
6 DLU3LC	0.003849	0.64	17	1.	6 DLU3LC	0.003847	2.65	1.
7 DLU3	0.000834	0.14	17	1.	7 DLU3	0.000834	0.57	1.
8 VRLU3	0.093121	15.57	18	1.35	8 VRLU3	0.012803	8.82	9.8
9 VRLU3L	0.124007	21.07	18	1.38	9 VRLU3L	0.021239	14.63	9.35
*** TOTAL	0.598138			1.35	*** TOTAL	0.145195		5.56
*** OVERHEAD	0.004578				*** OVERHEAD	0.004565		

vectorial "f" code (vect.off)   vectorial "f" (vect.on)

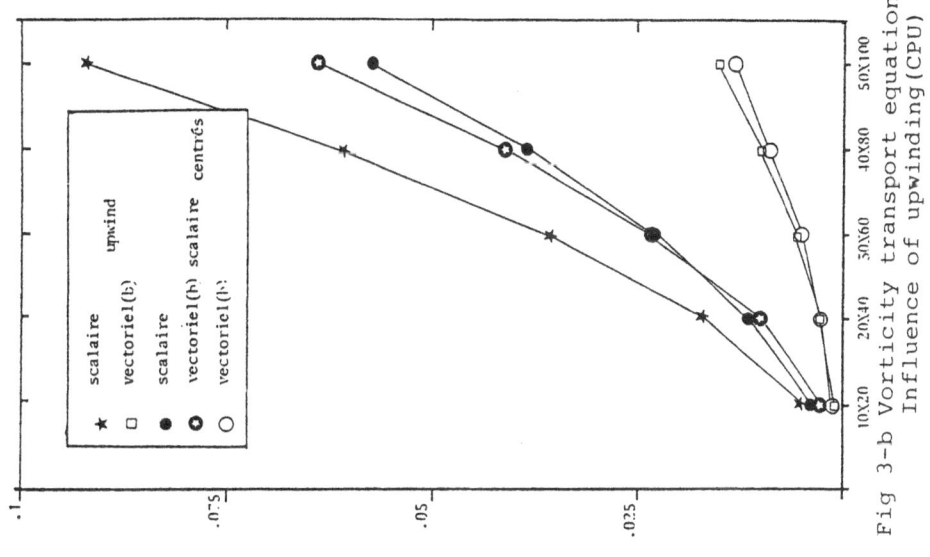

Fig 3-a Vorticity transport equation
Comparisons between codes (CPU)

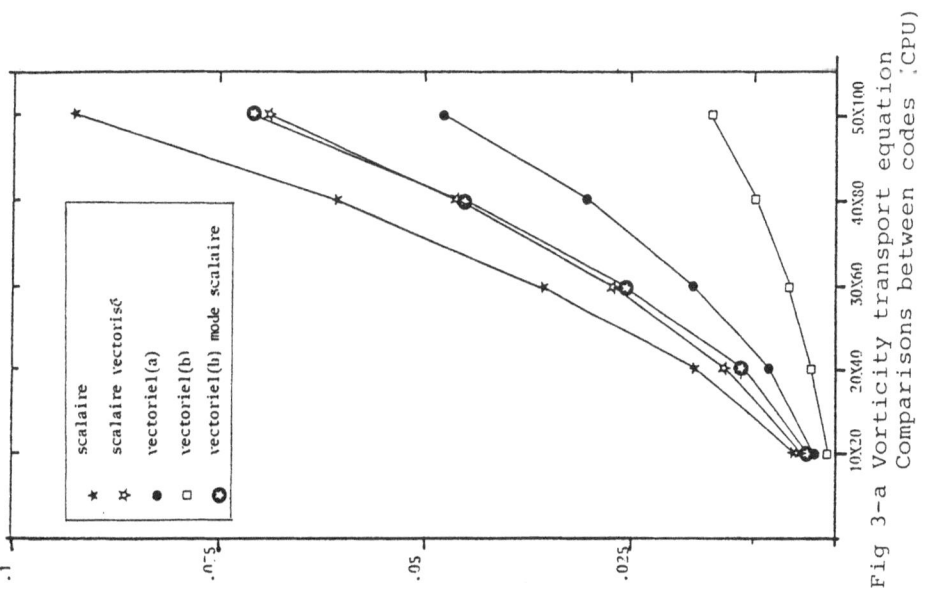

Fig 3-b Vorticity transport equation
Influence of upwinding (CPU)

167

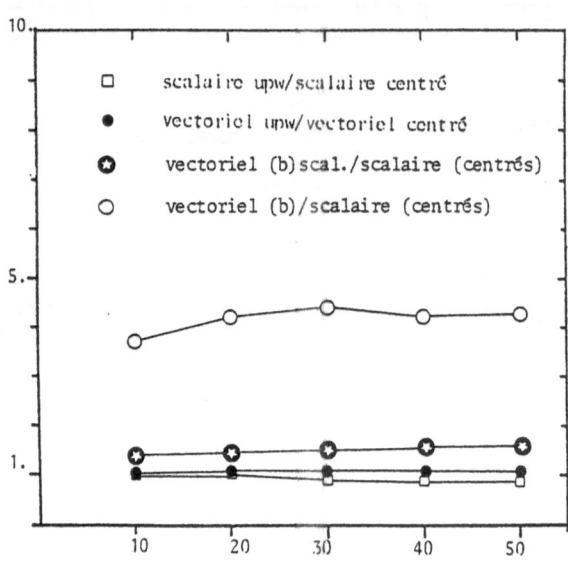

Fig. 3-c Vorticity transport equation
          Speed up ratios

Table 4 Vorticity transport equation; CPU time.

Essai	Maillage	10 x 20		20 x 40		30 x 60		40 x 80		50 x 100	
scalaire	(u)	$.5213 \ 10^{-2}$	1.	$.1701 \ 10^{-1}$	1.	$.3550 \ 10^{-1}$	1.	$.6067 \ 10^{-1}$	1.	$.9252 \ 10^{-1}$	1.
scalaire vectorisé	(u)	$.4441 \ 10^{-2}$	1.17	$.1356 \ 10^{-1}$	1.25	$.2730 \ 10^{-1}$	1.30	$.4589 \ 10^{-1}$	1.32	$.6912 \ 10^{-1}$	1.34
vectoriel (a)	(u)	$.2598 \ 10^{-2}$	2.	$.8362 \ 10^{-2}$	2.03	$.1752 \ 10^{-1}$	2.03	$.3062 \ 10^{-1}$	1.98	$.4667 \ 10^{-1}$	1.98
vectoriel (b) scal.	(u)	$.3571 \ 10^{-2}$	1.46	$.1146 \ 10^{-1}$	1.48	$.2557 \ 10^{-1}$	1.39	$.4532 \ 10^{-1}$	1.34	$.7073 \ 10^{-1}$	1.31
vectoriel (b)	(u)	$.1148 \ 10^{-2}$	4.54	$.3007 \ 10^{-2}$	5.66	$.5830 \ 10^{-2}$	6.10	$.1013 \ 10^{-1}$	5.99	$.1500 \ 10^{-1}$	6.17
scalaire	(c)	$.3905 \ 10^{-2}$	1.	$.1152 \ 10^{-1}$	1.	$.2297 \ 10^{-1}$	1.	$.3840 \ 10^{-1}$	1.	$.5750 \ 10^{-1}$	1.
vectoriel (b) scal.	(c)	$.2723 \ 10^{-2}$	1.42	$.1034 \ 10^{-1}$	1.12	$.2320 \ 10^{-1}$	.99	$.4106 \ 10^{-1}$	.96	$.6427 \ 10^{-1}$	.89
vectoriel (b)	(c)	$.1034 \ 10^{-2}$	3.77	$.2713 \ 10^{-2}$	4.25	$.5234 \ 10^{-2}$	4.40	$.9044 \ 10^{-2}$	4.24	$.1336 \ 10^{-1}$	4.30

Table 5 Vorticity transport equation; speed-up ratios

Maillages		10 x 20	20 x 40	30 x 60	40 x 80	50 x 100
scal./scal.vect.	(u)	1.17	1.25	1.30	1.32	1.34
vect.(b)scal./vect.(b)	(u)	3.11	3.81	4.38	4.47	4.72
vect.(b)scal./vect.(b)	(c)	2.63	3.81	4.43	4.54	4.81
scal.(u)/scal.(c)	(c)	1.33	1.47	1.55	1.58	1.61
vect.(b)(u)/vect.(b)(c)	(c)	1.11	1.11	1.11	1.12	1.12

done).

H⁴ Schemes :

With Mehrstellen schemes used with a Beam-Warming factorization the
vectorization of the computation of the matrices coefficients is easily
vectorizable because the schemes are centered; unfortunately,there is a
severe Reynolds cell limitation and it is impossible to increase the Reynolds number of the flow,as mentionned in [7]. With UOCI schemes the limitation is not so severe but the computation of the coefficient is difficult to
vectorize because a lot of tests are needed to insure some constraint for
each value of the index of the coefficients. The extensive use of vectorizable tests allows us to reduce the CPU time in the subroutines VOCI and
VOCI1;the results are presented in Table 6 for a classical test with a
101X151 mesh. Two version of VOCI1 are tested (VOCI1 and VOCI2) and differ by
the fact that the use of CVMXX functions are or not isolated in special DO
loops;it is important to notice that the multiplication of the number of DO
loops is efficient in vectorial code but less efficient for scalar mode. In
this case,the vectorial code without vectorizer is less efficient than the
original scalar routines. The two last rows of this table give the CPU-times
for Mehrstellen schemes and the ratio between the scalar and the vectorized
version of the subroutine . It is important to notice that the Mehrstellen
schemes are roughly 3.5 times faster than the U.O.C.I. schemes both in
scalar or vectorial mode.

Table 6 H⁴ schemes;vectorization of "VOCI"

subroutine	time	ratio
VOCI scal	3.1523	1.
VOCI vect	2.4737	1.274
VOCI1 scal	3.9875	0.791
VOCI2 scal	4.0308	0.782
VOCI1 vect	0.9361	3.367
VOCI2 vect	0.9168	3.438
MEHR⁴ scal	1.095	1.
MEHR⁴ vect	.2634	4.160

IMPULSIVELY STARTING CYLINDER

Some global results are now presented about the problem of the impulsive
motion of a circular cylinder;the value of the Reynolds number based on the

170

Table 7 Global test; Flow trace

FLOW TRACE --- SUMMARY — original scalar code (a)

ROUTINE	TIME	%	CALLED
1 VCYLCNS	0.382435	1.06	1
2 DPOICL	18.322387	50.76	11
3 DADIOP	0.001543	0.00	1
4 DTRI	0.000271	0.00	2
5 DLU3LC	0.506695	1.40	1507
6 DLU3	0.216572	0.60	1517
7 DRLU3	5.529631	15.32	27150
8 DRLU3L	9.087185	25.17	26969
9 DCOPIV	0.044246	0.12	714
10 DTRBA2	1.903335	5.27	10
11 DSIMP	0.003683	0.01	48
12 RESCYL	0.008184	0.02	2
13 LOCEXT	0.090746	0.25	2
*** TOTAL	36.096913		
*** OVERHEAD	1.722138		

FLOW TRACE --- SUMMARY — vectorized scalar code (b)

ROUTINE	TIME	%	G
1 VCYLCNS	0.119752	0.43	3.19
2 DPOICL	11.407003	41.02	1.6?
3 DADIOP	0.001512	0.01	1.
4 DTRI	0.000273	0.00	1.
5 DLU3LC	0.506480	1.82	1.
6 DLU3	0.216539	0.78	1.
7 DRLU3	5.528741	19.88	1.
8 DRLU3L	9.084606	32.67	1.
9 DCOPIV	0.008726	0.03	5.1
10 DTRBA2	0.834122	3.00	2.28
11 DSIMP	0.001124	0.00	3.27
12 RESCYL	0.007186	0.03	1.14
13 LOCEXT	0.090716	0.33	1.
*** TOTAL	27.806778		1.30
*** OVERHEAD	1.720909		

original scalar code (a)   vectorized scalar code (b)

FLOW TRACE --- SUMMARY — vectorial code (vect.off) (c)

ROUTINE	TIME	%	CALLED	G
1 VCYLCNS	0.379856	1.45	1	1.
2 VPOICL	13.158764	50.29	11	1.39
3 DADIOP	0.001542	0.01	1	1.
4 DTRI	0.000271	0.00	2	1.
5 DLU3LC	0.005769	0.02	17	1.46
6 DLU3	0.002428	0.01	17	.97
7 VTRB2	1.773321	6.78	10	1.07
8 MLU3LC	0.340846	1.30	10	
9 MRLU3L	0.513418	1.96	10	1.55
10 DCOPIV	0.043973	0.17	710	1.?
11 MLU3	0.721554	0.85	10	
12 MRLU3	0.320989	1.23	10	1.23
13 VRLU3L	5.304791	20.27	170	
14 VRLU3	3.946878	15.27	170	
15 DSIMP	0.003681	0.01	48	1.
16 RESCYL	0.008535	0.03	2	1.
17 LOCEXT	0.090755	0.35	2	1.
18 DROOT	0.000028	0.00	4	
*** TOTAL	26.167200			1.33
*** OVERHEAD	0.038123			

FLOW TRACE --- SUMMARY — vectorial code (vect.on) (d)

ROUTINE	TIME	%	G
1 VCYLCNS	0.119221	2.71	3.21
2 VPOICL	2.282870	51.88	8.08
3 DADIOP	0.001526	0.03	1.
4 DTRI	0.000271	0.01	1.
5 DLU3LC	0.005769	0.13	6.98
6 DLU3	0.002428	0.06	5.91
7 VTRB2	0.431942	9.82	4.4
8 MLU3LC	0.066812	1.52	
9 MRLU3L	0.078527	1.78	11.5?
10 DCOPIV	0.008662	0.20	5.1
11 MLU3	0.034202	0.78	
12 MRLU3	0.046166	1.05	9.88
13 VRLU3L	0.709460	16.12	
14 VRLU3	0.513605	11.67	
15 DSIMP	0.001124	0.03	3.27
16 RESCYL	0.007331	0.17	1.16
17 LOCEXT	0.090717	2.06	1.
18 DROOT	0.000028	0.00	
*** TOTAL	4.400663		8.2
*** OVERHEAD	0.038080		

vectorial code (vect.off)   vectorial code (vect.on)

(c)   (d)

diameter is 3000 and the mesh is 151X151. The vorticity equation is solved with help of the A.D.I. factorization with centered schemes in space,the number of iterations for Poisson equation is limited to 16 and the number of time step is 10.Here we test only the original code and the more efficient version of the vectorizable code with and without the vectorizer;the following table gives the references of these codes.

	vect.off	vect.on
"scalar code"	[a]	[b]
"vectorizable code"	[c]	[d]

All the results obtained are given in Table 7 and the global speed-up ratio [a]/[d] is 8.2 and the maximum ratio is got in the resolution of the systems of equations M and N. For exemple the routines (MRLU3L+VRLU3L) are 11.53 much faster than DRLU3L and (MRLU3+VRLU3) are 9.88 faster than DLU3.

COMPARISONS BETWEEN TWO COMPUTERS

In this section, some results are presented for the case of an impulsive starting cylinder; the original scalar code and the "vectorizable" one were tested on two computers, the NAS9080 and The CRAY1-s . They are summarized in the following Table 8.

Table 8 Comparisons between two computers

Code	Time	ratio1 NAS	ratio2 CRAY
Scalar	34.600	1.	—
Scalar	36.097	0.959	1.
Scal. Vect.on	27.807	1.244	1.298
Vect. Vect.off	26.167	1.322	1.379
Vect. Vect.on	4.401	7.862	8.202

The first conclusion is that in scalar mode, the NAS9080 seems to be a little faster than the CRAY1-S , but the accuracy is not the same because the lengh of the words is not the same ; for a real case, with output, the vectorial version of the code is globally 8.2 faster than the scalar one , which is a good result. The code tested is not fully optimized but the main work was done with the aims given at the beginning of this paper.

Acknowledgements

Computers facilities have been provided by the Scientific Committe of "Centre de Calcul Vectoriel pour la Recherche" . Financial support of DRET through Contract 83/473 is also gratefully acknowledged.

REFERENCES

[1] LECOINTE, Y. ,PIQUET J. "Compact Finite Differences Methods for Solving Incompressible Navier Stokes Equations around Oscillating Bodies". Von Karman Lecture Series 1985-04. Computational Fluid Dynamics (128 pages).

[2] COLLATZ, L. "The Numerical Treatment of Differential Equations " Springer Verlag 1966.

[3] KRAUSE, E. ;HIRSCHEL, E.H. ;KORDULLA, W. "Fourth Order "Mehrstellen" Integration for three-dimensional turbulent Boundary Layers" Comp. & Fluids 4 pp.77-92 1976.

[4] BEAM, R.M. ;WARMING, R.F. "Alternating Direction Implicit methods for parabolic equations with mixed derivatives ". SIAM J.Sci.Stat.Comp. 1 ;pp.151-159 1980.

[5] TA PHUOC, L. ;DAUBE, O. "A Mixed Compact Hermitian Method for the Numerical study of Unsteady Viscous Flow around an Oscillating Airfoil ". Proc. 3rd GAMM Conf. Num. Meth. Fluid Dyn., Notes Num. Fluid Mech. 2;pp 56-61 ; Ed E.H. HIRSCHEL (Vieweg Verlag) 1981.

[6] BERGER, A.E. ;SOLOMON, J.M. ;CIMENT, M. ;LEVENTHAL, S.H. and WEINBERG, B.C. "On generalized O.C.I. schemes for boundary layer problem " Maths of Comp. ,35, 155 pp. 79-94 1981.

[7] LECOINTE, Y. ; PIQUET, J. "On the Use of Several Compact Methods for the Study of Unsteady Incompressible Flows around Circular Cylinders". Comp. and Fluids Vol 12,4, pp.255-280 1984.

ON THE EFFICIENT USE OF LARGE DATA BASES IN THE
NUMERICAL SOLUTION OF THE NAVIER-STOKES EQUATIONS
ON A CRAY COMPUTER

W. Kordulla

DFVLR, Institut für Theoretische Strömungsmechanik
Bunsenstraße 10, D-3400 Göttingen, Germany

SUMMARY

For the numerical integration of the Navier-Stokes equations with the explicit-implicit MacCormack algorithm the efficient use of a CRAY-1S computer is
described, if the main memory is too small to handle the tackled problem. The
plane concept is used for the transfer of data. Within each plane of data the
code is nearly completely vectorized. To be really efficient dedicated
input/output devices are needed. Parallel input/output streams are used to
speed up the transfer rates. The set up of the code allows to handle large
problems, provided sufficient memory is available in core, and if dedicated
input/output devices are employed. The code is not well suited for efficient
use on a CYBER vector computer because the vector strings are comparatively
short.

INTRODUCTION

The advent of faster and faster supercomputers combined with improved and
computer-adapted numerical algorithms makes it more and more feasible to simulate the flow past three-dimensional configurations based on field solutions
of the governing equations. Loosely speaking, this development started in the
early seventies with the assumption of potential flow to reduce the number of
dependent variables to a single, scalar quantity, the velocity potential. In
the early seventies as well, the numerical integration of the time-dependent
and in particular steady-state Euler equations was initiated in applications
of research type. At about the same time first experiences were gained with
the numerical integration of the Navier-Stokes equations for compressible,
two-dimensional, high Reynolds number flows past e.g. airfoils, supersonic
compression corners or for shock wave boundary layer interaction problems.
All these investigations had to be carried out on serial computers such as a
CDC 7600 or even on slower machines. The investigations were, in particular in

the case of three-dimensional time-dependent cases, restricted to the choice of fairly coarse meshes as well as, often, not fully converged solutions. A one-of-a-kind computer with parallel architecture, the ILLIAC IV, on the other hand, enabled to simulate even the structure of (simple) three-dimensional turbulent flows. This, however, could only be achieved by adapting the coding of the computer program to the structure of the machine and in particular to the special organization of the memory. In the meantime computers have evolved with architectures that can handle most efficiently operations involving long strings of data, i.e. vector computers. Among these computers machines like CDC CYBER 205 require very long vectors and an appreciable amount of recoding [1], while machines like CRAY-1 or -XMP need only moderately long strings and comparatively few changes in the codes in order to operate efficiently. On such machines the simulation of two-dimensional viscous flows does not represent a problem from the numerical point of view with respect to space requirements or computation time (i.e. concerning the mesh resolution and time for convergence). In spite of the difficulties associated with the necessary storage space and computation time for computations in three dimensions, some attempts have been made recently which involve large, mostly out-of-core data bases. A fine mesh (roughly 1 million cells) is reported by Rizzi to have been studied for the integration of the Euler equations for the flow past three-dimensional wings on a CYBER 205 with sufficient in-core memory of 32 million words of 32 bits length. Deiwert and Rothmund report on the simulation of viscous three-dimensional flows on another CYBER 205 using more than 200000 mesh points in their investigation. A major portion of their work consists of coping with the difficulty that the fast main memory of the computer is much too small to contain all the data necessary for the computation. The same is true for present-day CRAY computers. Misegades reports to have used 2.5 million cells for the Euler solution for the transonic flow past a Dillner wing.

This paper reports on experiences, mainly on a CRAY-1S, gathered when integrating the time-dependent Navier-Stokes equations for compressible three-dimensional flows and large data bases. The physical problem is the transonic flow past a hemisphere-cylinder at angle of attack. Some physical and numerical aspects of the flow computations have been discussed in [2,3], while this paper is discussing the computational aspects in detail. The numerical scheme is MacCormack's explicit-implicit algorithm [4,5] which is vectorized for use on a CRAY-1S and which is based on a finite-volume formulation. The numerical scheme is displayed in [5], so that only those portions will be repeated which are needed to explain the computational procedures on the CRAY.

While the initially used mesh with 31 × 20 × 31 cells [2] fits entirely into the main memory of 1 million words of which roughly 855000 words can be used (no reduction to single precision words is possible), this is not the case with the 64 × 38 × 40 grid employed lately [3]. Using a 100 × 38 × 60 mesh allows to investigate the influence of the grid on the solution. Two concepts for input/output operations are known to the author. One is the pencil concept as used recently by Deiwert and Rothmund to arrive at long vector strings. This approach is very efficient for thin-layer Navier-Stokes computations where the turbulent viscosity coefficient is needed only for that last sweep through the data for which the boundary-layer like viscous terms are considered.

When, however, all viscous terms are retained in the Navier-Stokes equations, the corresponding expressions are needed from the very beginning. In this case the plane concept becomes useful in order to avoid additional sweeps through the data bases. Therein a (variable) number of computational planes, corresponding to wall normal surfaces in physical space, are contained in the core of the computer, such that all terms in the governing equations can readily be evaluated. Since the complete terms are used, the vector lengths cannot be made very long. This, however, does not impose serious penalties on a CRAY computer.

The need of out-of-core data bases may lead to increased turn-around times if dedicated input/output (I/O) devices are lacking. This is due to the limited speed with which data can be tranferred between the computer mainframe and the storage device, typically disks. In such cases the computation must wait for data to be written on disk, and for data to be read from disk, in exchange for the buffered-out data. This I/O wait time may be many times larger than the actual computation time (CPU time).

In the following, those features of the numerical algorithm will be discussed first which influence the computational performance on vector computers. Then the plane concept, the heart of the out-of-core version, is explained. And finally the used input/output procedures will be discussed, and the computational performance on the CRAY-1S will be compared with that one on a CRAY-XMP4/8 where just one processor is being used, without changing the coding.

REQUIREMENTS OF THE EXPLICIT-IMPLICIT MACCORMACK SCHEME

First, the equations to be integrated numerically, based on a finite-volume formulation of the explicit-implicit MacCormack scheme, are briefly given as a basis of the discussion. The integral form of the governing equations form the basis of the explicit integration steps:

$$\frac{\partial}{\partial \tau} \int_V q(g)^{1/2} \, d\upsilon + \oint_{\partial V} (\underline{q}\underline{v}+\underline{b}) \cdot \underline{n} \, dS = 0 \quad, \tag{1}$$

$$\underline{q} = (\rho, \rho\underline{u}, \rho E)^T \quad, \quad \underline{v} = \underline{u} - \underline{u}_{mesh} \quad,$$

$$E = f_{pe} \frac{T}{\gamma - 1} + f_c \frac{1}{2}(\underline{u})^2 \quad,$$

$$\underline{b} = (\underline{b}_\rho, \underline{b}_m, \underline{b}_e)^T \quad,$$

$$\underline{b}_\rho = 0 \quad, \quad \underline{b}_m = f_{pu} \, p\underline{I} + Re^{-1}\underline{\underline{T}} \quad,$$

$$\underline{\underline{T}} = -\lambda \mathrm{div}\,\underline{u}\,\underline{I} - \mu[(\mathrm{grad}\,\underline{u})+(\mathrm{grad}\,\underline{u})^T] \quad,$$

$$\underline{b}_e = -\gamma(RePr)^{-1}\mu\,\mathrm{grad}\,T + f_{pe} \, p\underline{u} + f_c Re^{-1}\underline{\underline{T}}\,\underline{u} \quad.$$

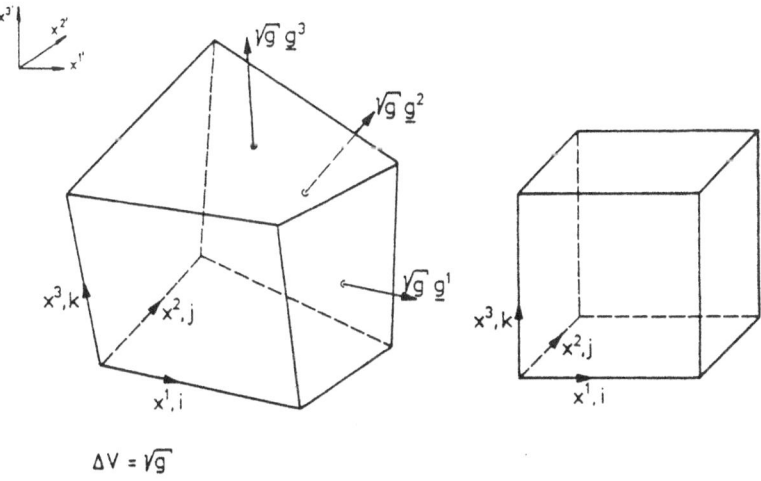

$$\Delta V = \sqrt{g}$$

Figure 1: Sketch of the control volume in physical and computational space.

Here, all terms are arbitrarily non-dimensionalized with the corresponding freestream quantities except for the speed of sound which is referenced with the freestream velocity [5]. For convenience, a transformation is assumed so that a structured mesh is implied, see figure 1. The term $(g)^{1/2}$ is the volume of the considered control volume, and it is. identical to the inverse of the Jacobian which is usually used in finite-difference approaches. Equations (1) read in discretized form:

$$\Delta [\underline{U}(g)^{1/2}] + \Delta\tau \sum_{l=1}^{3} \Delta_l \left\{ [q\underline{v}+\underline{b}] \cdot (g)^{1/2} g^l \right\} / \Delta x^l = 0 , \qquad (2)$$

where the terms emanating from the surface integrals are approximated in a one-sided fashion in each of the predictor and corrector steps so that after the complete time step some average will be taken of the fluxes contributed by the cells of both sides of the surface in question. The implicit steps can best be viewed as an approximation of the partial differential equations obtained by differentiating the differential form of the governing equations with respect to time. The differential form corresponding to equations (1) is given by

$$\frac{\partial}{\partial\tau} [\underline{q}(g)^{1/2}] + \frac{\partial}{\partial x^1} [(q\underline{v}+\underline{b}) \cdot (g)^{1/2} g^1] = 0$$

or

$$\underline{\hat{q}},_\tau + {}^1\underline{\hat{F}},_1 = 0 .$$

After having been differentiated with respect to time this equation becomes:

$$\underline{\tilde{\hat{q}}},_\tau + ({}^1\underline{\hat{A}}\,\underline{\tilde{\hat{q}}}),_1 = 0 , \quad \underline{\tilde{\hat{q}}} \equiv \underline{\hat{q}},_\tau , \quad {}^1\underline{\hat{A}} \equiv \partial\,{}^1\underline{\hat{F}}/\partial\underline{\hat{q}} . \qquad (3)$$

The discretized form of equations (3) after factorizing the implicit operator reveals a product of bi-diagonal, one dimensional operators. Choosing the direction x^3 as an example the bi-diagonal implicit equation reads:

$$[\underline{I} + \Delta\tau^3\underline{C}_k] \delta\underline{\bar{U}}_k = \delta\underline{U}_k^{**} + \Delta\tau^3\underline{C}_{k+1} \delta\underline{\bar{U}}_{k+1} , \qquad (4)$$

where the bi-diagonal algorithm is obvious. It is also obvious that the implicit inversion becomes obsolete it the coefficient matrix on the left hand side vanishes. This is the case if some one-dimensional stability condition is satisfied [4,5].

178

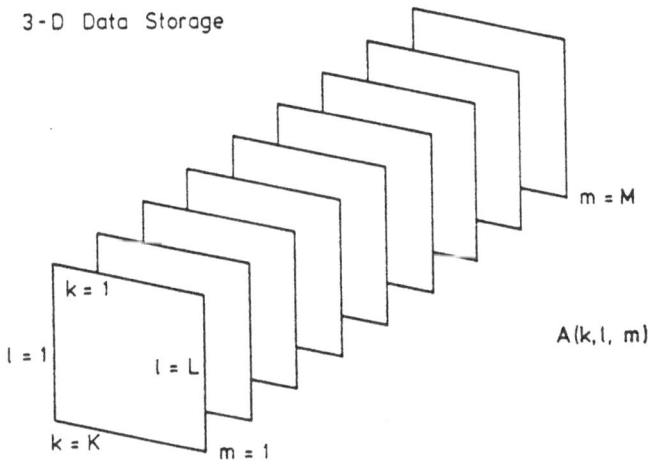

3-D Data Storage

$k = 1$

$l = 1$ $l = L$

$k = K$ $m = 1$

$m = M$

$A(k, l, m)$

Figure 2:   Arrangement of three-dimensional arrays in the computer storage.

Cell i

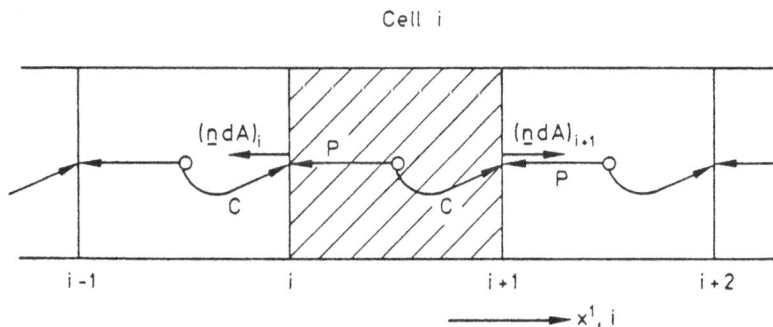

$(\underline{n}dA)_i$ P $(\underline{n}dA)_{i+1}$

P

C   C

$i-1$ i $i+1$ $i+2$

x^1, i

Typical Predictor - Corrector Sequence

Figure 3:   Sketch of the surface-flux relationships in the predictor (P) and
corrector (C) sweep.

The complete explicit-implicit algorithm is as follows:

Predictor:

$$\Delta \underline{U}^n = -\Delta\tau (\Delta_+^{\ 1}\hat{\underline{F}}/\Delta x^1) \ , \quad 1 = 1(1)3 \ ,$$

$$(L_1 L_2 L_3)_+ \ CFL_2 \ \delta\bar{\underline{U}} \ = \ CFL_2 \Delta\underline{U}^n \ , \tag{5a}$$

$$CFL_1 \equiv \min\,[\,1.0, 0.5\,CFL/CFL_{i,j,k}\,] \ , \quad CFL_2 \equiv 1. -CFL_1 \ ,$$

$$\bar{\underline{U}} \ = \ \underline{U}^n + CFL_2 \ \delta\bar{\underline{U}} + CFL_1 \ \Delta\underline{U}^n \ .$$

Corrector:

$$\Delta\bar{\underline{U}} \ \ = -\Delta\tau (\Delta_-^{\ 1}\bar{\hat{\underline{F}}}/\Delta x^1) \ , \quad 1 = 1(1)3 \ ,$$

$$\overline{(L_1 L_2 L_3)}_- \ \overline{CFL_2} \ \delta\underline{U} \ = \ \overline{CFL_2} \ \Delta\bar{\underline{U}} \ , \tag{5b}$$

$$\overline{CFL_1} \ = \min\,[\,1.0, 0.5\,CFL/\overline{CFL}_{i,j,k}\,] \ , \quad \overline{CFL_2} \equiv 1. - \overline{CFL_1} \ ,$$

$$\underline{U}^{n+1} \ = \frac{1}{2}\,[\,\underline{U}^n + \bar{\underline{U}} + \overline{CFL_2}\ \delta\underline{U} + \overline{CFL_1}\Delta\bar{\underline{U}}\,]\ .$$

Here, the factor L_i denotes the implicit bi-diagonal operator with respect to the direction x^i which results in a difference equation corresponding to equation (4), where i = 3.

The explicit predictor and corrector steps by themselves are one-sided approximations. It is well known that this introduces a bias into the solution, which vanishes only for constant coefficient matrices (Jacobians) at the end of the predictor-corrector sequences. In general, symmetry in the operators and therefore in the solution can only be achieved by interchanging the predictor and corrector sequences appropriately in successive time steps. The same properties apply to the implicit operations. In figure 2 the arrangement of data in the computer storage is being displayed. In the case of the explicit integration sweeps the need for symmetric operations does not impose any difficulties with respect to the data management. This need is satisfied by considering the control surface as main element. In the predictor and in the corrector sequence, the proper flux is then obtained by using the variables of the appropriate cell next to the surface in question, see figure 3. It then suffices to sweep through the data always in only one direction, say from i = 1 to i = IL, see figure 3. Equation (4), on the other hand, shows that unidirectional sweeps through the data are not sufficient for the implicit steps because on the right hand (explicit) side of the equations the unknown flux appears as boundary condition, either up- or downstream of the cell in question. Hence, for the bi-diagonal operator one has to sweep in different directions depending whether forward or backward differences are being used. This bears consequences with respect to the data management because one desires to choose the sweeping direction in a more less arbitrary way as needed.

The CFL factoring of the solution in equations (5) has some disturbing effect as far as storage requirements are concerned. For the in-core version of the code it can be avoided to introduce additional arrays to perform the CFL factorization of the explicit side, one portion of which is needed at the end of the predictor and corrector step, respectively. Instead the two three-dimensional arrays necessary for the integration process, one for the known time-level or last step and one for the new time-level or current step, are stored away and overwritten by using asynchronous BUFFER-IN and -OUT procedures on sequential files. Because the multi-dimensional arrays are one order of magnitude smaller in the out-of-core version it was decided to use one additional array to reduce the I/O procedures as much as possible. Three solution arrays are used, in fact. The additional array allows to read the known solution at the end of the predictor-corrector sequences, and, otherwise, takes care of the CFL-factored explicit side.

THE PLANE AND PENCIL CONCEPT FOR DATA TRANSFER

As soon as the fast in-core storage of the used computer becomes too small for the problem to be solved, one has to look for an efficient way to move data between the in-core memory and out-of-core storage devices. The way depends, of course, on the kind of supercomputer at one's disposition, and on the chosen mathematical and physical approach.

In [6] a CYBER 205 is used which works the more efficient the longer the vector strings are. On the other hand the thin-layer approximation of the Navier-Stokes equations is integrated, in which the classical boundary layer assumption is used for the approximation of the viscous terms, so that these contain just derivatives normal to the wall of the considered (axisymmetric) body. Due to the factorization of the implicit operator the viscous terms and therefore the corresponding transport coefficient are needed only in the wall normal solution sweep. The expressions for the eddy viscosity, used usually in turbulence modeling, can hence be collected in the course of the sweeps through the data base such that during the last sweep the viscous terms are finally computed. This feature is exploited to produce large vector strings in [6]. By dividing the computational data planes in squares, see for example figure 4 for the x^i-direction, Deiwert und Rothmund achieved vector lengths

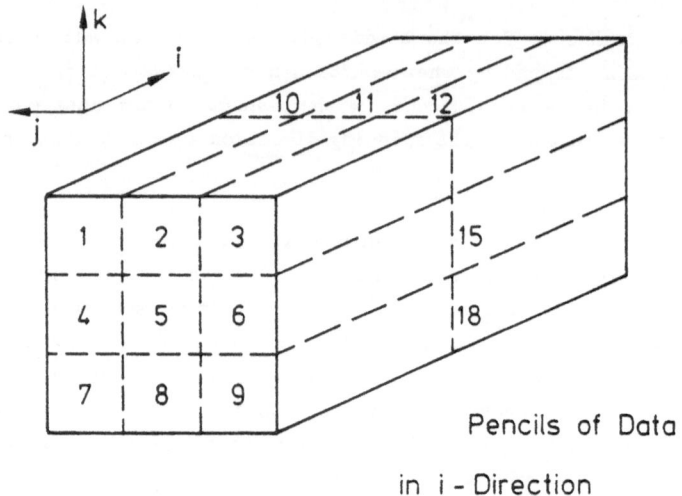

Pencils of Data

in i - Direction

Figure 4: Sketch of the partitioning of the date base for the pencil concept.

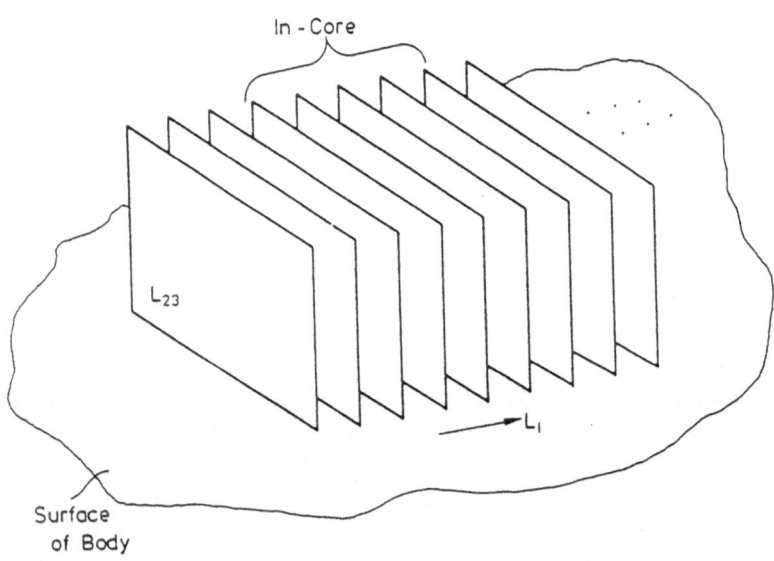

Figure 5: Sketch of the partitioning of the data base for the plane
concept.

of 20 × 20 = 400 which produce a reasonable efficiency on a CYBER 205. The structure of the partitioning is pencil like in the sweeping direction and hence its notation 'pencil concept'.

If the full Navier-Stokes equations were used, the pencil concept is not attractive because of the derivatives to be approximated within the computational plane. Also, for the determination of the eddy viscosity, two additional sweeps through the data base would be needed, so that the viscous terms can be computed from the very beginning on. This can be avoided without additional data handling if the plane concept is adopted, where at all times just a certain number of data planes are kept in core of the computer, see figure 5. In the current version of the code this number is four. The data planes in question are chosen such that the corresponding surfaces in physical space are nearly normal to the wall in the immediate neighbourhood of that wall, in order to possibly justify a thin-layer approximation. There is just one main sweeping direction, namely that one normal to the data planes. All operations can be performed in the course of the sweep. For central schemes and first-order accurate upwind (or "flux-split") schemes those two planes at the ends of the in-core block act as boundary planes, and the solution is determined for the remaining in-core data planes, see figure 6. For the second-order accurate upwind scheme, see e.g. [5], two boundary planes on each end are required away from the boundaries of the overall computational domain, and a minimum of five planes is needed in core. The solution, however, is computed just in one plane. The requirement of a minimum of five in-core planes can reduce the maximum number of grid points within each of the in-core data planes considerably if the main memory is small (< 1 million words). For the former schemes the minimum number of planes is three. If four planes are used the solution is computed in two which increases the efficiency with respect to I/O in comparison with the case with three in-core planes. A draw back of the plane concept is the relatively short length of the vector strings (say e.g. 18 to to 36 in the current example). Therefore, the plane concept is advocated for use on CRAY but not on CYBER computers.

COPING WITH THE I/O BOTTLENECK ON A CRAY-1S

The in-core version of the code has been vectorized nearly completely [7] by using mostly the auto-vectorizing features of the CRAY-1S computer. The resulting performance is at least competitive with that one of other, comparable codes [7]. The vectorization implies that the rules given in the CRAY computer manuals are being followed such as:

• make innermost DO-loops most efficient, i.e. vectorizable and longest,

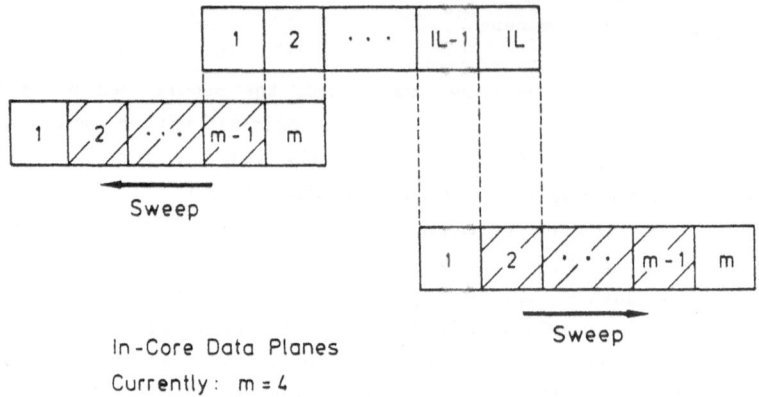

In-Core Data Planes

Currently: m = 4

Figure 6: Sketch of exchange of the in-core planes.

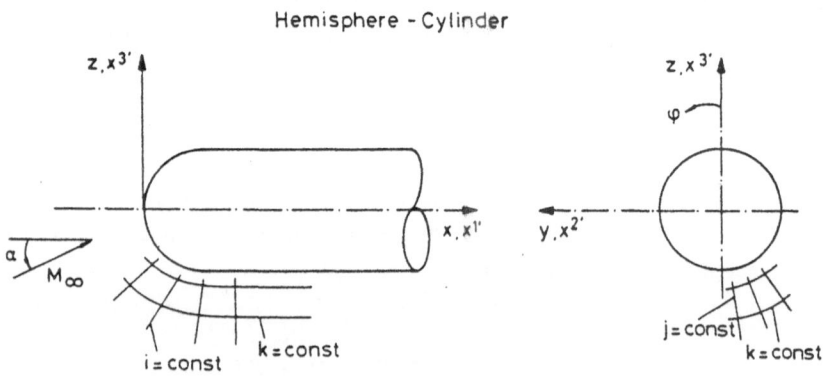

Figure 7: Sketch of the hemisphere-cylinder configuration.

- unroll short DO-loops,

- remore IF- and CALL-statements,

- replace IF-statement by Conditional-Vector-Merge Function,

- decompose non-vectorizable DO-loops into shorter vectorizable ones,

- use vectorizable DO-loops for strings of points which are treated in an implicit way in contrast to the local, point-wise handling in the serial code.

The only unvectorized portions of the code concern the determination of the strings of points where the integration is carried out in an implicit way. By direct comparison with the original serial code it was found that the number of implicitly treated cells does not increase considerably by forming strings of cells instead of the local checking and inversion. In addition to the mentioned changes of the serial code, the unformatted I/O operations have been replaced by asynchronous BUFFER-IN and -OUT statements for sequential files. This cut the I/O times down to half of the original times, but was not important with respect to the overall performance since the I/O uses just a negligible portion of the overall computation time. This situation changes dramatically for the out-of-core version with large data bases.

The physical example for the simulation of the flow field is the transonic flow past a semi-infinite hemisphere-cylinder combination at angle of attack, see figure 7. The maximum number of cells for the in-core version in i,j and k direction, see figure 7, was 42 × 20 × 31. Since a grid with fewer cells, 32 × 20 × 31, did not give a different answer the latter is chosen for comparison purpose in comparison with the out-of-core version. The in-core version used roughly 24 variables contained in the following arrays:

2 solution vectors:
    VA1 (i,j,k,5)
    VA2 (i,j,k,5)

19 metric quantities
(coordinates of cell corners, surfaces, volume):
    XYZ (i+1,j+1,k+1,3)
    SI (i+1,j+1,k+1,3)
    SJ (i+1,j+1,k+1,3)
    SK (i+1,j+1,k+1,3)
    VOL (i,j,k)

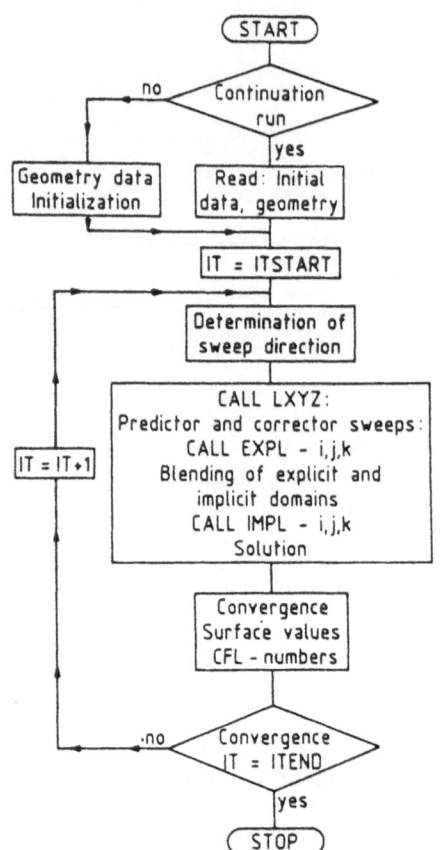

Figure 8: Sketch of the structure of the in-core version of the code.

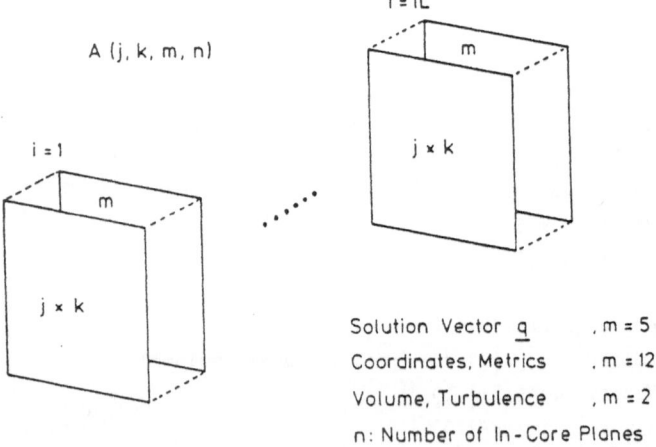

A (j, k, m, n)

i = 1

i = IL

j × k

m

j × k

m

Solution Vector q   , m = 5
Coordinates, Metrics   , m = 12
Volume, Turbulence   , m = 2
n: Number of In-Core Planes

Figure 9: Sketch of the arrangement of the stored data for the out-of-core version.

turbulence model (eddy viscosity):

VIST (i,j,k).

Because it was judged more efficient with respect to the overall effort, the metric quantities (e.g. surface normals SI, SJ and SK) are kept in main memory and are not recomputed each time these quantities are needed. Typical CPU times per cell and time steps for a slightly implicit version (note that the check whether the solution is locally implicit or not requires considerable work in each of the three directions, and is always performed) are $9.3 \cdot 10^{-5}$ seconds for the full Navier-Stokes equations, while the thin-layer approximation and the Euler equations require 6 and $5.1 \cdot 10^{-5}$ seconds, respectively. These numbers are valid for laminar flows. These times include the time spent in subroutines to do some data reduction such as computing surface values and CFL numbers. In the turbulent case, using the well-known Baldwin-Lomax model, 10^{-5} seconds has to be added. The structure of the code is given in figure 8. This is not changed in the out-of-core version. There, however, the procedure indicated in figure 8 is carried out just for those data planes which are in core. And an outer DO-loop takes care of the proper plane wise march through the data base. The out-of-core version is applied to cases with grids with $64 \times 38 \times 40$ ($\simeq 85000$) and $100 \times 38 \times 60$ ($\simeq 205000$) cells. In order to achieve efficient I/O operations [1]) essentially 29 variables were used in the following arrays (n: number of in-core planes)

2 solution vectors:

VA1 (40,64,5,n),
VA2 (40,64,5,n),

1 dummy solution vector:

VA3 (40,64,5,n),

12 metric quantities:

XYZ (40,64,12,n+1),

2 quantities (volume, eddy viscosity):

VOLT (40,64,2,n).

Characteristic is the size of the data plane, 40,64 which corresponds to five blocks (1 block = 512) of words. This is because the data to be moved must be a

---

[1]) The author gratefully acknowledges the help of H. Cornelius and J. Pichlmeier, CRAY Research Inc., W. Germany, in streamlining the performance of the out-of-core version.

multiple of blocks for unblocked I/O. Figure 9 displays how the four-dimensional arrays are being stored.

The out-of-core version was conceived because at the time no supercomputer was around with sufficiently large main memory. And still today and in the near future only a few institutions can afford to have one of those bigger supercomputers which commence to appear in the market. The idea then is to retain the vectorization performance, described for the in-core version, for the data planes in core and to use some efficient I/O strategy to exchange the in-core data.

In that case the ideal situation for a CRAY user is a dedicated I/O device such as the Solid-state Storage Device (SSD) with high-speed-channel connection to the mainframe (100 to 1000 million bytes per second compared with an optimal 10 million for disks). The author, however, has to live with the features of a CRAY-1S without I/O Subsystem and without SSD. The main storage memory contains 1 million words with about 855000 usable ones.

The requirement to perform arbitrary sweeps through the data base leads to the use of direct-access files. CALL GETPOS and CALL SETPOS procedures are chosen for random direct-access files. Thus the first word of any of the arrays shown in figure 9 can be addressed. Furthermore, asynchronous BUFFER-IN and -OUT is used to enable parallel execution of code and I/O, if possible, and also I/O itself in parallel. The operating system in that case requires the use of unblocked files, i.e. multiples of 512 words have to be transferred (hence the use of arrays like A(64,40,m,n), see above). In order to really exploit the asynchronous features multiple-stream I/O is being used. Unfortunately code execution can not occur while I/O operations are being performed. Three different disk controllers are employed to transmit data to and from three different disks, in parallel. Then for a $100 \times 38 \times 60$ (205000) cells grid the rate of data processing, as defined by the CPU time per cell and time step in 10^{-5} seconds, is 9 on a CRAY-1S and 5.3 on a CRAY-XMP/48 using just one processor (the latter data have been obtained by H. Cornelius, CRAY Research Inc.). With the older machine the ratio of waiting time for I/O and CPU was generally between 1.25 and 1.75 depending on the work load of the computer. This ratio reduced to 1 on the XMP without using the SSD while the ratio became negligible ($\simeq 0.007$) as soon as the data files were put on the SSD. It should be mentioned that the ratio of waiting time for I/O and CPU time can easily obtain values such as 13 if the programming and data management is not done carefully. In the case of four in-core data planes, for example, two of the data blocks shown in figure 9, have to be moved in and out whenever another step is taken in the i-direction. Initially, this was done in a sequential

188

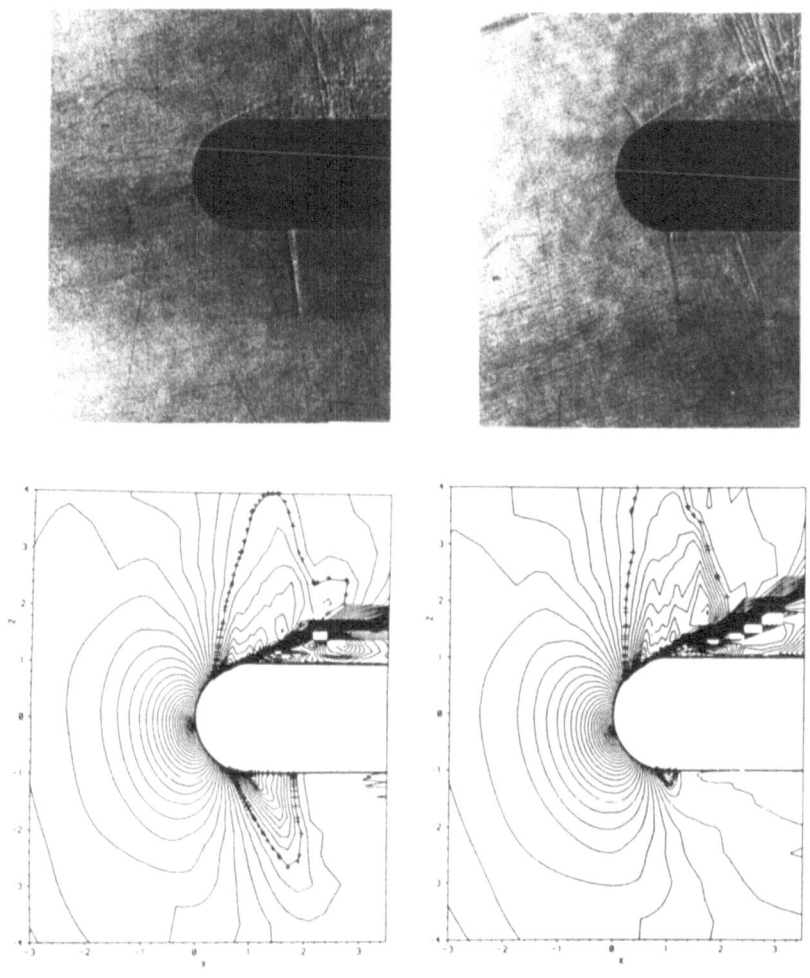

Figure 10:   Comparison of shadowgraph pictures with the computed distribution
             of lines of constant Mach number (+: sonic lines),
             $M_\infty = 0.9$, Re = 212500, $\alpha = 10°$ (left) and $19°$ (right).

way plane by plane while BUFFER-IN and -OUT was used. By extending the BUFFER operations over the two blocks in one step a tremendous reduction in waiting time for I/O was obtained. An example of the solution obtained with an intermediate mesh of 64 × 38 × 40 cells is presented in figure 10. The freestream Mach number is 0.9, the Reynolds number based on the radius 212500 and the angle of attack is 10° and 19°. Shadowgraph pictures of the flow in the symmetry plane are compared with lines of constant Mach number. In spite of the asumption of laminar flow the overall agreement, in particular near the nose, is very good: the separating shear layer in the leeward symmetry plane as well as the shock in the forward symmetry plane for 10°.

CONCLUSIONS

It has been shown that the vectorized in-core version of a code, based on the finite-volume, explicit-implicit MacCormack scheme, has been extended to an out-of-core version using the plane concept, without loss of efficiency other than that due to the lack of dedicated I/O devices. It is believed that the presented solution concept can in principle handle any number of data planes in the plane marching direction. The resolution of the flow field in the plane itself, however, is limited by the magnitude of the main memory. The solution, shown in figure 10, uses 38 × 60 cells in the circumferential direction with the assumption of the existence of a symmetry plane. Thereby about 614000 out of 855000 words of memory are used. Thus, the number of cells could be increased somewhat, in particular if the number of in-core data planes was reduced to three. However, it is believed that 855000 words of memory are insufficient to allow for a solution, which features asymmetric flow, without sacrifying the necessary circumferential mesh resolution.

REFERENCES

[1] GENTZSCH, W.: Vectorization of Computer Programs with the Application to Computational Fluid Dynamics. Vieweg-Verlag, Notes on Numerical Fluid Mechanics, Vol.8, 1984.

[2] KORDULLA, W.: The computation of three-dimensional transonic flows with an explicit-implicit methods. Proceedings, 5th GAMM-Conference on Numerical Methods in Fluid Mechanics, M.Pandolfi/R.Piva (Eds.), Notes on Numerical Fluid Mechanics, Vol.7, pp.193-202, Vieweg-Verlag, 1984.

[3] KORDULLA, W.: The Computation of Three-Dimensional Viscous Transonic Flows with Separation. Proceedings, 9th ICNMFD, ed. Soubbaramayer, Lecture Notes in Physics, Springer-Verlag, 1984.

[4] MacCORMACK, R.W.: A Numerical Method for Solving the Equations of Compressible Viscous Flow, AIAA Journal, Vol.20, pp.1275-1281, 1982.

[5] KORDULLA, W. and MacCORMACK, R.W.: A New Predictor-Corrector Scheme for the Simulation of Three-Dimensional Compressible Flows with Separation. AIAA 85-1502, 1985.

[6] DEIWERT, G.S. and ROTHMUND, H.: Three-Dimensional Flow Over a Conical Afterbody Containing a Centered Propulsive Jet: A Numerical Simulation. AIAA Paper 83-1709, 1983.

[7] KORDULLA, W.: MacCormack's Methods and Vectorization. In: W. Gentzsch, Vectorization of Computer Programs with the Application to Computational Fluid Dynamics; Vieweg-Verlag, Notes on Numerical Fluid Mechanics, Vol.8, pp. 157-171, 1984.

N3S : A 3D FINITE ELEMENT CODE FOR FLUID MECHANICS; EMPHASIS TO VECTORIZATION AND SPARSE MATRIX PROBLEMS.

Ph. HEMMERICH, J. GOUSSEBAILE, J.P. GREGOIRE, P. LASBLEIZ

ELECTRICITE de FRANCE
Département Mécanique et Modèles Numériques,
Département Laboratoire National d'Hydraulique

I   -   INTRODUCTION

N3S is a three-dimensional incompressible Navier-Stokes finite element code developped at EDF for the study of industrial flows[*]. The treatment of complex geometries, the number of nodes (up to 50000) led us to decompose the problem at different levels (geometry, algorithm and numeric). The use of the splitting up technique allows a specific and efficient treatment to be applied to each part of the Navier-Stokes equations :

- method of characteristics for the advection terms,
- conjugate gradient method for the diffusion-continuity part.

The most recent N3S development is a $k-\varepsilon$ turbulence model with wall function and thermal effect. (first results will be presented in [5]).

The N3S code runs on a CRAY-1 vector computer. Some of the ideas of this code are hardly connected to the performances of such a computer (slow I/0, disk limits and vector capability). We present here some results on the matrix vector product and the preconditionning in the conjugate gradient, with regard to the vectorization capability.

---

[*]   (Pipes, reactor core, turbo-machine for internal flows and cooling tower, small atmospheric effects for external flows).

II - DECOMPOSITION OF THE NAVIER-STOKES EQUATIONS : fractionnal steps method

The non-linear advective terms of the NS equations are decoupled using a fractionnal step algorithm :

$$\frac{\partial u}{\partial t} + u\nabla u - \nu\Delta u + \frac{1}{\rho}\nabla p = F \qquad \text{+ boundary conditions}$$

$$\frac{\partial \tilde{u}}{\partial t} + u^n\nabla\tilde{u} = 0 \qquad \text{advection equation} \qquad (1)$$

$$\frac{u^{n+1}-\tilde{u}}{\Delta t} - \nu\Delta u^{n+1} + \frac{1}{\rho}\nabla p^{n+1} = F \qquad \text{Stoke's problem} \qquad (2)$$

$$\text{div } u^{n+1} = 0$$

The basic idea is to make advective terms explicit by using an upwinding scheme as physically as possible, and to build an accurate scheme of high order which guarantees unconditional stability (no time-step limitation) with low numerical diffusion. This is obtained by upwinding along the physical characteristics curves defined by :

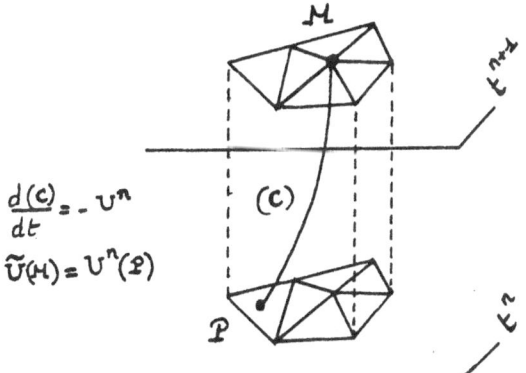

$$\frac{d(c)}{dt} = -U^n$$

$$\tilde{U}(M) = U^n(P)$$

The computation of that curve is solved by applying a Runge-Kutta method of order 2.

The classical Galerkin finite element discretization of the Stoke's problem leads to the following system of equations :

$$AU - {}^t BP = S \qquad (S = F + \frac{1}{\Delta t}\tilde{U})$$

$$- BU = 0 \qquad (3)$$

where A is a symmetric positive definite matrix and B is the rectangular matrix resulting from the discretization of the divergence operator. The global symmetric system (3) could be solved using a LL^T decomposition but in order to decrease on one hand the CPU time, on the other hand the main storage we prefer an iterative solution :

The Uzawa algorithm

It can be interpreted as a gradient method for the pressure equation :
$$\left[\, ^{t}BA^{-1}B \, \right] \quad P \quad = \, - \left[\, ^{t}BA^{-1} \right] \, S \qquad \text{obtained by elimination of the velocity in (3)}$$

$$P = P_0$$

Until $\| B \, u^n \| \leq \varepsilon$ Preconditionned Uzawa algorithm

$$\left| \begin{array}{rcl} A u^n & = & S + {}^{t}B \, P^n \\ C R^n & = & B \, u^n \\ P^{n+1} & = & P^n - \rho_n \, R^n \end{array} \right.$$

The main advantages of this method are :

- separate problems on the pressure and <u>each</u> component of the velocity which are uncoupled, with Dirichlet boundary conditions, in the matrix A.
- great decreasing of storage due to the possibility of systematic packing for the matrix A.

<u>Remark</u> : each problem $AU = S + {}^{t}BP$ is solved either by a LL^T factorization (with a sky-line storage) or by a conjugate gradient (with a packing storage)

III - <u>SKY-LINE OR PACKING STORAGE</u> ?

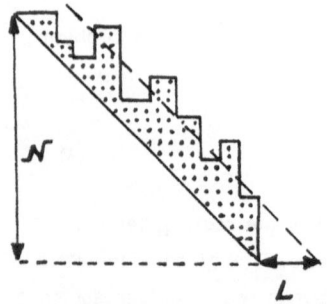

For the factorization of $A = LL^T$, the theoretical CPU time is equal to :

TCPU $= k \, L \, (NL)$ operations
where L is the mean bandwith
NL is the number of non-zero coefficients
kL is the number of operations to compute one coefficient.

So that TCPU = $k \, NL^2$

In pratice we have established from some computationnal tests that the effect of vectorization on the CRAY-1 leads to the following expression :
TCPU $\simeq$ k' (NL^2) $^{0.75}$ sec. with k' = 3.05 10^{-6} .

If we extrapolate this result with a very simple mesh, made of 3D quadratic (Q2) bricks with m bricks in each direction :

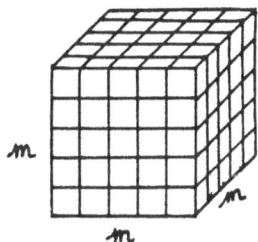

m	bricks Q2 (m^3)	nodes $(2m+1)^3$	L	TCPU	Storage
5	125	1.331	480	7s	0.6 Mword
10	1.000	9.261	1750	210s	16 Mwords
15	3.375	29,791	3800	27mn	113 Mwords
20	8.000	68,920	6700	2h	460 Mwords

Clearly the packing storage is more suitable if we consider the number of non-zero coefficients of the matrix as proportional to the number of nodes (the number of non-zero coefficient is nearly constant for each row of the matrix, with a finite element or difference discretization) - The conjugate gradient (CG) method is an efficient and well-adapted method to solve a linear symmetric and positive definite system Ax = b with a standard packing storage. The drawback of packing storage is the non-regular storage of the data and then a low degree of vectorization on some computers like the CRAY-1.

IV - VECTORIZATION AND PRECONDITIONNED CONJUGATE GRADIENT (PCG)

Problem : solve $Ax = b$

Initialization : x^0 given .

Iteration k , $r^k = Ax^k - b$ is the residual vector

$$Mz^k = r^k \qquad\qquad M : \text{preconditionning matrix}$$

$$\gamma = (r^k, z^k) / (r^{k-1}, z^{k-1})$$

$$d^k = z^k + \gamma\, d^{k-1} \qquad\qquad \text{new direction}$$

$$\rho = -(r^k, d^k) / (Ad^k, d^k)$$

$$x^{k+1} = x^k + \rho d^k \qquad\qquad \text{new residual vector}$$

$$r^{k+1} = r^k + \rho\, Ad^k \qquad\qquad \text{new iteration.}$$

test on $\| r^{k+1} \|$.

The system $Mz = r$ has to be "easily" solved, for instance :

$M = I$

$M = $ diagonal of A

$M = LL^T$ incomplete Cholesky factorization (ICF) .

For each iteration we have to :

    solve the system $Mz = r$

    multiply the matrix A with a vector d

    compute 3 dot products

    compute 3 linear combinations of vectors

    compute a vector norm.

If we choose $M = $ Identity or M diagonal of A, then 95% of the CPU time is spend to compute the product A.d.

4.1 **Matrix-vector product,** A is a packed matrix.

We suppose the matrix A stored with a standard packing mode, let say (IDIAG, IP, AC) where

IDIAG =		N - vector
IDIAG(I)=		adress of the diagonal coefficient for the ith column in the vector AC
AC	=	NC - vector , NC is the number of non-zero coefficients of A
IP	=	NC - vector
IP(I)	=	index of the row for the ith coefficient of AC

Then the product A.d may be computed with the following routine :

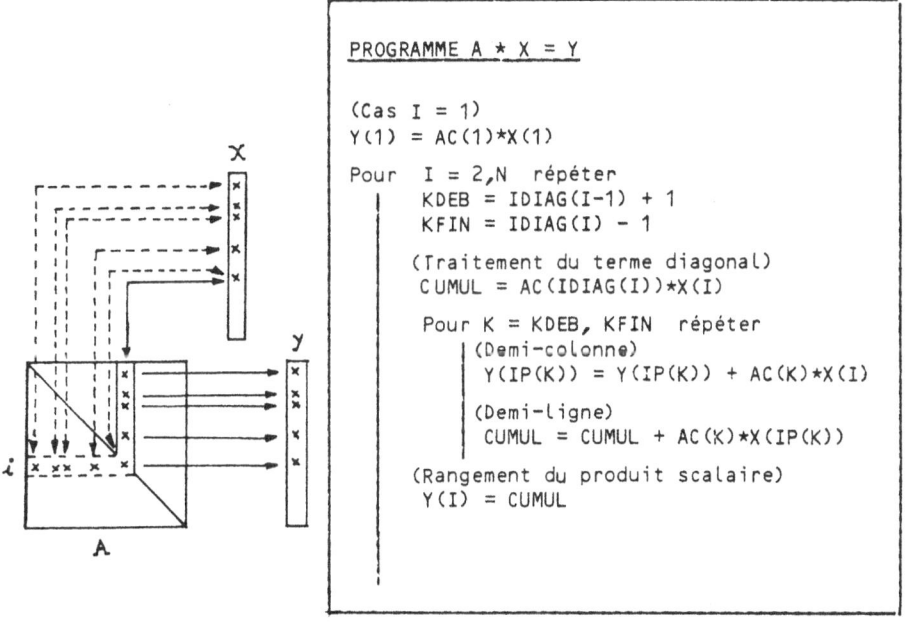

```
PROGRAMME A * X = Y

(Cas I = 1)
Y(1) = AC(1)*X(1)

Pour  I = 2,N  répéter
   KDEB = IDIAG(I-1) + 1
   KFIN = IDIAG(I) - 1

   (Traitement du terme diagonal)
   CUMUL = AC(IDIAG(I))*X(I)

   Pour K = KDEB, KFIN  répéter
      (Demi-colonne)
      Y(IP(K)) = Y(IP(K)) + AC(K)*X(I)

      (Demi-ligne)
      CUMUL = CUMUL + AC(K)*X(IP(K))

   (Rangement du produit scalaire)
   Y(I) = CUMUL
```

Problem : The indirect adressing, in other words the non-regular storage of the data, prevents the most inner-loop from being vectorized on the CRAY-1, unless we use the GATHER and SCATTER routines. But the size of the inner loop remains small (between 10 and 50) and the price to pay with the Gather or Scatter is important as we can see on the following results.

For each line the first number is the CPU time, the second one (with parenthesis) is the ratio to the first vector loop case. For all values of N, the third loop is better than the second one whereas that's not true for small N between the 5th and the 4th loops.

N $\longrightarrow$	10	50	100	1000	10000
Vector DO I = 1,N Y(I) = Y(I) + A(I) * X(I)	1.7 (1.)	4. (1.)	7.5 (1.)	69. (1.)	685 (1.)
Scalar DO I = 1,N Y(I) = Y(I) + A(I) * X(IP(I))	6.5 (3.8)	30. (7.5)	60. (8.)	600. (8.7)	6000. (8.8)
CALL GATHER ( ) DO I = 1,N Y(I) = Y(I) + A(I) * XX(I)	6.4 (3.8)	13. (3.3)	20. (2.7)	170. (2.5)	1600. (2.3)
Scalar DO I = 1,N Y(IP(I)) = Y(IP(I)) + A(I) * X(I)	5.5 (3.2)	25. (6.3)	50. (6.7)	500. (7.2)	5100. (7.4)
CALL GATHER ( ) DO I = 1,N XX(I) = XX(I) + A(I) * X(I) CALL SCATTER ( )	10.5 (6.2)	22. (5.5)	33. (4.4)	263. (3.8)	2570. (3.8)

1.
To
3.8

0.5
To
2.

we are here !

Front storage : In order to take advantage of vector computers we must exchange the short inner loop and the outer loop of size N (=several thousands), that is to say we must replace the row or column storage by a front storage of order N (we call front a set of index so that two consecutive coefficients of the front belong to two consecutive rows of the matrix)

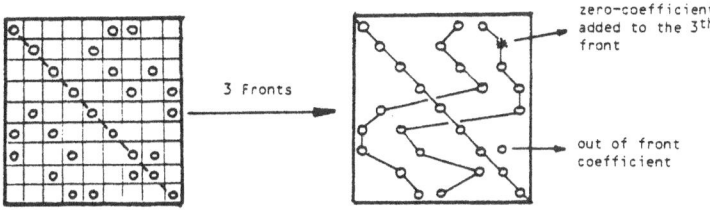

The drawbacks of this kind of storage are :

-   unsymmetric storage
-   some zero-coefficients are added into the fronts
    some coefficients remain stored out of the fronts and have to be treated separatly
-   what is the optimal number of fronts ?

A "good" matrix must have already the same number of coefficients on each row; that is true with a finite element discretization on linear (Q1) bricks, but not on quadratic (Q2) bricks. In accordance with that we can save 50% of time with Q1 bricks but only 20% in the case of Q2 bricks.

Choice of the nodes numbering

One advantage of packed matrices is to give an amount of storage which is independent of numbering; so we don't care of the renumbering problem. Nevertheless it is possible to increase the performances using a renumbering such that the number of stored zero-coefficients will be minimized : sort the nodes considering the number of their neighbours and choose first the most to other nodes connected nodes; the different sizes of the fronts allow us to suppress an important number of unuseful operations (multiplication with zero) and can save 45% of CPU time in the case of Q2 bricks.

Next we have performed some tests with regular numbering coming from grid mesh (or finite difference mesh) without holes. In this case of course the product A.d may be vectorized without gathering or scattering the data, so that it becomes easy to save 85% (Q1) or 75% (Q2).

4.2 Preconditionning of the CG

The classical ICF choice for the preconditionning matrix M leads to solve two triangular linear systems for each iteration of the CG

$$Mz = LL^T z = r , \quad Lu = r , \quad L^T z = u .$$

L being sparse and packed as well as the matrix A, the same problem of indirect addressing occurs and induces the same sanction : inhibition of the vectorization. Furthermore the cost of IC preconditionner takes 65% of the time from the whole CG. Then we have to improve precontionning rather than the product of a sparse matrix with a vector.

The main problem arises from the difficulty to make up one's wind with regard to the efficiency of a preconditionning; it depends not only on the topological structure of the matrix (hence on the numbering) but on the value of its coeffients (hence on the physical problem).

Yet a very simple idea seems to give same hopful results : to keep from the ICF only the coeffients lying inside a fixed width band along the diagonal. The triangular systems solver becomes easy to vectorize (but with the order of the bandwith).

Remark : This method works well only if the coefficients of the matrix A are mostly closed to the diagonal, therefore after a renumbering algorithm like Cuthill-Mackee's.

If we go further why not to take an approached inverse $\tilde{L}$ of the triangular band matrix L, keeping the same bandwith ? The preconditionning problem is then very simple to solve, it's the product of a band matrix with a vector, a high speed operation on a vector computer.

First numerical test have been performed with the N3S code for the three u, v and w (4128,3344 and 3902 degrees of freedom) matrices resulting from a physical problem (cooling tower with a wind incoming).

To explain the quite low number of iterations required for each CG, it is a time dependent approach and each one of the linear system Ax = b has been solved with an initial vector x° equal to the solution at the previous time step, hence closed to the vector solution.

Precondi-tionner	Number of iterations	A.d time	Mz = r time	Global time	Ratio
IC	4	156 ms	280 ms	436 ms	1.
No	22.2	866 ms	-	866 ms	2.
Diagonal	9.7	378 ms	5 ms	385 ms	0.9
11 Diag.	7.25	283 ms	191 ms	474 ms	1.1
22 Diag.	6.4	250 ms	191 ms	441 ms	1
22 Diag. L inver-sed	6.5	253 ms	51 ms	304 ms	0.7

Remark : (The choice of 22 diagonals gave in this case the same amount of storage for the $\tilde{L}$ - matrix than for the IC L-matrix). Inspite of an increasing of 60% of the mean number of iterations required for each CG, the inversed approched $\tilde{L}$ matrix remains the best choice with regard to the global CPU time. What's more with this method 80% of the time is spent for the product A.d in scalar mode, which can be improved with some work.

CONCLUSION

Some results with regard to the product of a sparse matrix and a vector are disappointing : especially with irregular matrices coming from a Q2 discretization (the benefit of the vectorization gets only to 20% or to 45% with a front storage and an adequate renumbering). On the contrary the regular matrices are well adapted to a full vectorization, because they don't need a gathering and scattering of the data.

In order to make the finite element codes more competitive several solutions are to be considering in the future :

- to use hardware Gather.
  It is available with the CRAY-XMP or CYBER 205; that is of course the simplest way !
- to use domain decomposition.
  The idea is then to divide the computational domain into substructures in order to
  - reduce the computational cost by using the regularity of some parts of the whole demain (why not even with a finite difference method ?).
  - obtain independent problems for each subdomain, which is suitable for parallel computers.

BIBLIOGRAPHY

[1] J.P. BENQUE, P. ESPOSITO, G. LABADIE - New Decomposition Finite Element Methods for the Stokes Problem and the Navier-Stokes Equations -Numerical Methods in Laminar and Turbulent Flow (August 1983, Seattle, USA).

[2] J.P. GREGOIRE, J.P. BENQUE, P. LASBLEIZ, J. GOUSSEBAILE - 3D Industrial Flows Calculations by Finite Element Method - 9th International Conference on Numerical Methods in Fluid Dynamics (June 1984, Saclay, FRANCE).

[3] J.P. BENQUE, J.P. GREGOIRE, A. HAUGUEL, M. MAXANT - Application des méthodes de décomposition aux calculs numériques en hydraulique industrielle - 6ème Colloque International sur les méthodes de calculs Scientifiques et Techniques (Décembre 1983, Versailles, France).

[4] J. GOUSSEBAILE, J.P. GREGOIRE, A. HAUGUEL - Iterative Stokes Solvers and Splitting Technics for Industrial Flows - 5th International Symposium on Finite Element Methods in Flow Problems (January 1984, Austin, USA).

[5] J. GOUSSEBAILE, A. JACOMY, A. HAUGUEL, J.P. GREGOIRE - A Finite Element Algorithm for Turbulent Flow Processing a K-ε Model - Numerical Methods in Laminar and Turbulent Flow (July 1985, Swansea, U.K.).

[6] M.A. AJIZ, A. JENNINGS - A Robust Incomplete Choleski - Conjugate Gradient Algorithm - International Journal of Numerical Methods in Engineering (vol. 20, pp. 949-966, 1984).

[7] P. GILL, W. MURRAY, M. SAUNDERS, M. WRIGHT - Sparse Matrix Methods in Optimization - Siam Journal (vol. 5, n°3, September 1984).

MULTITASKING THE CODE ARC3D
John T. Barton
Christopher C. Hsiung

Abstract

The CFD code ARC3D has been run on a Cray X-MP/48 computer at the Cray Research, Inc. headquarters in Mendota Heights, Minnesota. The Cray multitasking system was invoked in order to utilize all four processors and sharply reduce the wall clock run time. This paper discusses the techniques used to modify the code for this run and analyzes the achieved speedup.

John T. Barton is the High Speed Processor System Manager with the NAS (Numerical Aerodynamic Simulation) Projects Office at the NASA Ames Research Center. Christopher C. Hsiung is with Cray Research, Inc., at Chippewa Falls, Wisconsin.

The ARC3D Code

The starting point for our work in multi-tasking was the computational fluid dynamics code ARC3D. This code solves either the Euler or Thin-Layer Navier Stokes equations using an implicit approximate factorization scheme. The earliest version of this code was written by Joseph Steger of the Ames Research Center. Later enchancements were added by Thomas Pulliam of Ames, and Jack Benek of AEDC in Tullahoma, Tennessee. The multi-tasking was implemented by Christopher Hsiung. The ARC3D code is in wide use as a benchmark program, in its earlier versions. The more current versions of the code continue to be used for algorithm development and computational physics of fluid flow. The most recent algorithmic enhancements to this version of ARC3D include a spatially variable time step, and a freestream stagnation enthalpy boundary condition. The spatially variable time step is a device to allow for more rapid convergence to a steady state solution; it cannot be used for a time varying calculation. We computed only steady states, selecting as our test case the flow over a hemisphere-cylinder. The boundary condition was justified because the stagnation enthalpy (enthalpy of a volume which has been isentropicly slowed to rest) should equal the stagnation enthalpy of the freestream flow. The density along the body was derived from this enthalpy requirement.

Hardware

The Cray X-MP/48 supercomputer is a four CPU (central processor unit) version of the popular X-MP line. The CPU clock time on the X-MP computers is 9.5 nanoseconds, 25 percent faster than the Cray-1 line. In addition, the Cray X-MP line features operation chaining and multiple ports to memory, which together with the faster CPU clock result in a significant performance improvement over the Cray-1.

In addition to the increased computational performance rate, the Cray X-MP/48 features a significantly larger main memory than either the Cray-1 line or earlier Cray X-MP models. The Cray X-MP/48 has a main memory of 8 million 64 bit words organized into 64 banks. Because the memory chips utilize a fast bipolar technology, the bank reservation time is only 4 CP ticks. This low value of the bank reservation time ensures that memory bank contention will be kept to a tolerable level even with four processors actively accessing memory.

System Software

The Fortran compiler used to compile the code was CFT (Cray Fortran) version 1.15. The Cray Operating System (COS) version 1.15 was the operating system. These versions of the compiler and the operating system are experimental and not yet available to Cray customer community. The Cray multitasking software system, which consists of both CFT Fortran constructs and COS system subroutines, was used as the tool for dividing the problem into separate tasks.

The Cray CFT compiler, especially versions 1.14 and 1.15, is quite effective in recognizing DO loops that are vectorizable and in generating the necessary vector machine instructions. As a result, only a few lines of code needed to be inserted or changed for all

205

significant inner loops to vectorize. Most of the lines inserted were compiler directives, in the form of special comment lines, to force vectorization on loops that were not immediately recognized as vectorizable by the compiler. Some other lines of code were changed, mainly to enhance the chaining of vector instructions and to avoid constructs that are not readily vectorized.

The Cray multitasking software has its roots in the IRTF (Industrial Real Time Fortran), a standard proposed by the European Workshop on Industrial Computer Systems in 1980. The first version available to the Cray user community utilized subroutine calls to the operating system to control and coordinate the multitask computation. The multitasking of the ARC3D code, however, was performed using a set of compiler directives that have been developed by one of the authors (Christopher Hsiung). This system of compiler directives, combined with a source code preprocessor, enable multitasking to be invoked without alterations to the Fortran code, since the directives are in the form of special comment lines. Examples of these compiler directives are as follows:

- CDIR$ TASK COMMON – Specifies that the COMMON block in the statement immediately following is local to each separate task.

- CDIR$ DO PARALLEL – Specifies that each iteration of the DO loop that begins on the next line may be performed in multiple streams.

- CDIR$ MONO PROCESS – Specifies that the code beginning with the next line until a corresponding END PROCESS directive must be executed only by the master process.

- CDIR$ MONITOR – Specifies that the code beginning with the next line until a corresponding END MONITOR directive is a critical section and must be executed by only one process at a time (usually because the code accesses some global data).

Modifications to the ARC3D code

The multitasking of the ARC3D code was performed at the outer DO loop level. That is, each iteration of the outer DO loop was performed as a separate task. Usually the outer DO loop was the index of one dimension of a key array. Assigning each value of the index as a separate task assured that the work would be divided as evenly as possible between tasks. Since each chosen index defined a plane in the computational grid, the parallel planes were handled on distinct processors.

Approximately 50 lines, or about 1% of the lines of code in the program file, were inserted or changed in order to invoke multitasking. This amount of revision is less than the amount usually required to optimize for the Cray X-MP a Fortran program that is otherwise intelligently written for a vector computer.

The ARC3D code has been run without multitasking on a Cray X-MP/22 computer at NASA Ames Research center with a 30 x 30 x 30 grid. The much larger memory available on the Cray X-MP/48 provided the opportunity to run the program using a 60 x 60 x 30 grid. This extended run required 4.5 million 64 bit words of main memory.

Speed-Up Analysis

In the following analysis, α will denote the degree of theoretical degree of parallelism in a code. In other words, α denotes the fraction of the run time that is potentially executable by concurrent processing. V will denote the multitasking overhead of the job. This includes the time spent in calls to initialize the multitasking environment, calls to start and synchronize the tasks, and memory contention. P denotes the number of processors and S_T denotes the theoretical speedup of multitask processing compared to single task processing. These parameters are related as follows:

$$S_T = \frac{1}{1 - \alpha + \alpha/P} \quad .$$

In the case where overhead associated with multitasking is included, which we denote by V, the expression is:

$$S_T = \frac{1}{1 - \alpha + (\alpha/P)(1 + V)} \quad .$$

For the ARC3D code, the degree of parallelism α is 0.988 for either a 30 x 30 x 30 grid or a 60 x 60 x 30 grid. Thus the theoretical speedup factor is 3.86. The actual speedup factors achieved in these two runs were 3.5 and 3.4, respectively. The performance rates for these runs were 233 and 279 MFLOPS (millions of floating point operations per wall clock second), respectively. The performance rates for the same jobs without multitasking were 66.8 and 82.5 MFLOPS, respectively.

In each case the actual speedup was only about 91% of the theoretical speedup. The discrepancy between the theoretical speedup and the actual observed speedup may have a number of sources. One possible source is the multitasking overhead – in other words the amount of time consumed by the multitasking control subroutines. This, however, is not suspected of being the major cause. A more likely source of this discrepancy is memory bank contention. The ARC3D code is highly CPU and memory intensive. In addition, a number of the important inner DO loops access arrays by other than the first dimension, which sometimes exacerbates memory bank contention problems. Thus it is likely that a significant amount of processing time was consumed waiting for main memory banks to be released from a previous memory access.

The overhead V for the 30 x 30 x 30 case was about 10.8%. The overhead for the 60 x 60 x 30 case was somewhat higher, being about 14.2%. This latter figure can be broken down into the portion due to load imbalance, which is 5%, and the portion largely due to memory contention, which was 9.2%. The load imbalance portion of the overhead is due to the fact that the loops of length 30 (in the 30 x 30 x 30 case) were broken up as follows over the processors: the end points were stripped off, and the remaining 28 points were equally divided among the 4 processors. In the 60 x 60 x 30 case, the relevant 58 points could not be equally divided, and the two processors which had 1 fewer point were idle during the last iteration of the other processors.

Conclusions

From our experience it appears that multitask processing can be used to achieve wall clock speedup factors of over three times, depending on the nature of the program code being used. Multitasking appears to be particularly advantageous for large memory problems running on multiple CPU computers. Such problems are most efficient when they can use most, if not all, of the available main memory. If they do not use more than one CPU, it is likely that the other CPUs would be idle while the large job is running, because it would be difficult to fit additional jobs in the remaining main memory. Thus multitasking represents a way to efficiently use the resources of a multiple CPU system such as the Cray X-MP/48 when running a job that requires large amounts of CPU time and main memory.

It should be noted, however, that at present it is rather difficult to invoke multitasking on a large scientific computation code, especially for someone that has not attempted such a conversion before. A major difficulty lies in debugging a computer run that is inherently asynchronous. Problems or errors that surface in one run might not recur the next time the job is run. Not helping the problem is that the software packages currently available for invoking multitask processing are generally difficult to use and frequently not fully debugged. In addition, constructs such as COMMON blocks in the Fortran language are inherently troublesome for concurrent computation, and sometimes subtle errors can arise, for example, when arrays in a common block are used as subroutine arguments. Finally, it is sometimes necessary to reorganize data structures in an existing code because the author did not have parallel processing in mind when writing the code. This problem should diminish as newer codes are written which have more parallel structure.

Despite the temporary difficulties of doing multitasking, which are the growing pains of the field, the work is important and needed. Supercomputers with multiple processors will soon be the rule, rather than a novelty. All the major supercomputer vendors will be manufacturing them; for instance the Cray-2 which the Numerical Aerodynamic Simulation Project will acquire in Fall of 1985 will have four processors, and the Cray-3 may well have 16. It is reasonable to assume that in the next century the number of processors will be bounded by the capability of software to deal efficiently with them.

THE EFFICIENT USE OF THE CRAY X-MP MULTIPROCESSOR VECTOR COMPUTER IN COMPUTATIONAL FLUID DYNAMICS

A.K. Dave
CRAY RESEARCH (UK) LTD, London Road
Bracknell, Berkshire, England

SUMMARY

For many problems in computational fluid dynamics the elapsed times for simulations are too large to make the routine use of the technique in research and development pragmatic, even with a vector computer such as the Cray-1. With the Cray X-MP series of supercomputers the elapsed time for such jobs may be reduced by more than a factor of seven. In this paper we discuss the factors important to the efficient exploitation of the novel features of these machines.

1. INTRODUCTION

Historically, the limits of computer technology have been a major impediment to the use of state-of-art techniques in computational fluid dynamics (CFD). The introduction of the Cray-1 supercomputer made it pragmatic to use many existing methods routinely and to develop more sophisticated models. Since then the complexity of the flows being modelled has increased substantially so that, even with such vector computers, the elapsed times for the most sophisticated modelling undertaken today are relatively large. The long elapsed times arise from the large amount of processing involved in the simulations and/or the fact that data cannot be held in central memory. The advent of multiprocessor vector computers, namely the Cray X-MP series, makes it possible to reduce the elapsed times for CPU-bound simulations dramatically. This may be achieved by multitasking the code to run on two, or more, processors simultaneously. Multitasking is an additional dimension in parallel processing on these machines which can be used when computing in both the serial and vector modes. A parallel development which addresses the problem of I/O bound codes is the evolution of the Cray Solid State Storage Device - a very large secondary memory coupled to the mainframe by a very fast channel.

In this paper we discuss the factors affecting the efficient mapping of simulation codes onto the Cray X-MP series architecture. As the concept of vector processing is relatively well known [1] we concentrate on the exploitation of the multitasking features. It is stressed, however, that vectorisation remains an important contribution to the processing speed. In section 2 the architecture of the Cray X-MP series and the roles of multitasking and vector processing are reviewed briefly. A laser fusion code, described in section 3, is used to illustrate several points. After discussing some constructs with large scale parallelism (section 4) we consider factors affecting multitasking efficiency. Section 6 presents the performance figures for multitasked codes.

2. CRAY X-MP ARCHITECTURE; VECTOR PROCESSING & MULTITASKING

In this section we first review the architecture of the Cray X-MP series of mainframes. We then discuss the roles of vector processing and multitasking. Multiple central processors are found in two members of the Cray X-MP range; the X-MP/2 with 2 CPUs and the X-MP/4 with 4 CPUs. Each of the CPUs is an enhanced version of the central processor of the Cray 1 vector computer. Fig.1 shows the organisation of a system based on a Cray X-MP/4 (fig 1). The mainframe communicates with the front end system through the input/out subsystem. The Cray X-MP multiprocessors have a closely coupled configuration in which the CPUs share a common, central, memory. Special hardware enables efficient multitasking. The processors may communicate with each other through shared data and semaphore registers (which monitor the use of shared resources) as well as the main memory. As each CPU is a vector processor we have to consider two aspects when large codes are mounted on these machines; vectorisation and multiprocessing.

The aim of multitasking is to reduce the elapsed time for a simulation. This is achieved by dividing the job into subtasks which may execute concurrently on separate CPUs. We note two circumstances in which multitasking is useful. Firstly, to reduce the time required for large CPU-bound problems, and secondly when the job uses most of the computers central memory. Cray Research has developed software to assist the multitasking of programs. This software simplifies the creation of tasks, the declaration of interdependencies and intertask communication. The utilities available for multitasking FORTRAN 77 programs are described in ref 4. With multitasking it is possible to exploit several levels of parallelism simultaneously.
The number of CPU cycles required are increased by multitasking due to the overhead of task creation and synchronisation. In contrast to this, vector processing increases the processing rate by reducing the number of CPU cycles required for execution. Multitasking and vector processing are complimentary dimensions of parallelism - whilst vector processing takes place at the level of the innermost Do loops, the devision of a job into subtasks should be as global as possible. Vectorisation typically reduces the processing time by factors between 10 and 20. Multitasking with p processors offers an additional speed-up $\leq$ p (p = 4 for the X-MP series). As a general guideline, the vector performance of the program should be optimised first. Multitasking should then be used to exploit parallelism at the highest level possible.

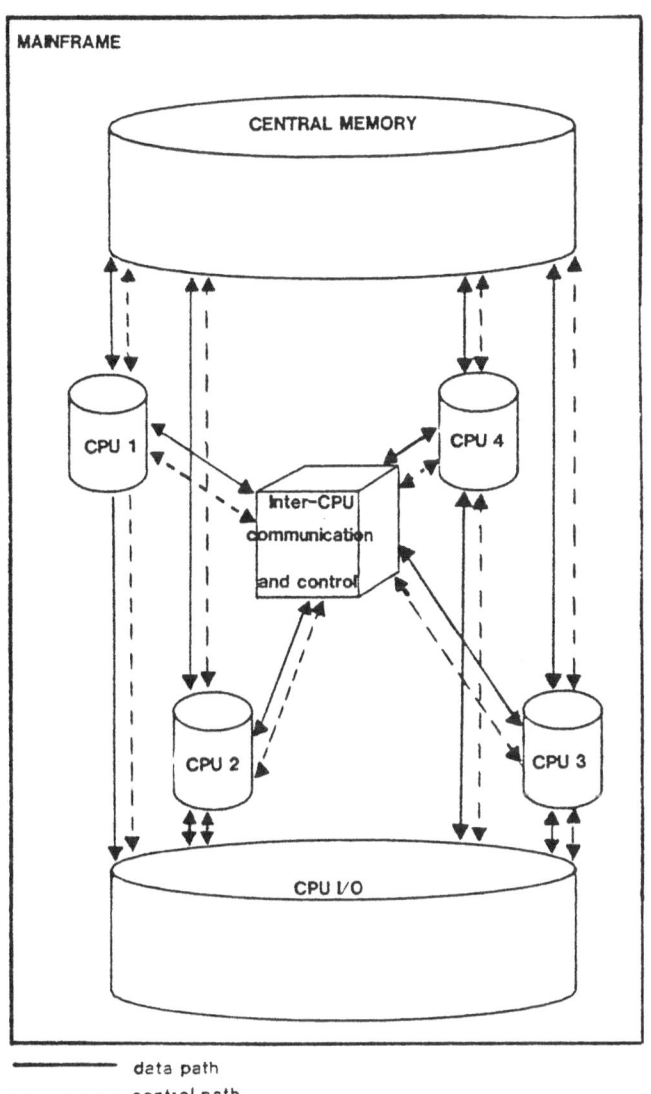

MAINFRAME

CENTRAL MEMORY

CPU 1

CPU 4

Inter–CPU
communication
and control

CPU 2

CPU 3

CPU I/O

———————— data path
-------- control path

Figure 1.   The system organization of the
            CRAY X-MP4.
            The data paths are depicted by solid
            lines and the control paths by bro-
            ken lines

3. DESCRIPTION OF POLXM

This section describes a laser fusion program developed at the University of Hull [2]. We have optimised the vector performance of this code and developed a multitasked version (MPOLXM) for execution of a Cray X-MP/2. In the rest of the paper we will use our experience in this exercise to illustrate several points.

POLXM is a finite difference code modelling the temporal development of plasmas produced by laser irradiation of matter. Representing the plasma as a quasi-neutral fluid, it evaluates the two dimensional hydrodynamics. Using the usual notation, the user may select his geometry from the following options: The (r,θ) spherical polar, (z,r) or (r,θ) cylindrical polar, and cartesian coordinates. The mesh is automatically rezoned to follow the fluid - a feature which improves the resolution of important structures in the flow. The user may select between quasi-lagrangian rezoning and a rezoning algorithm maintaining a uniform mesh. The mesh is constrained within specified boundaries and maintains an orthogonal structure. [Two fluids with classical energy exchange model the thermodynamical properties of the ions and electrons in the plasma.] The absorption and reflection of laser radiation is also simulated. The code is strongly conservative and solves the energy equation, including electron and ion thermal transport, using an implicit method. Figure 2 shows the structure of the code whose major components are described in more detail in the following sub-sections, which may be omitted at first reading.

3.1 Time Step Constraints (Sub:Time)

In addition to satisfying the usual Courant-Friedrichs-Lewy stability criterion for fluid codes, the time step is constrained to ensure the accuracy of the energy equation solution. This restriction ensures that the amount of energy conducted across a cell boundary or deposited by the laser light during a time step is small compared to the total thermal energy in the cell. A further restriction is applied to prevent the cells from emptying.

3.2 Co-efficient Calculation (Sub:Kintic)

The kinetic coefficient such as thermal conductivities are evaluated in kintic.

3.3 Attenuation of the laser beam (Sub:Absorb)

The deposition of the laser energy in the sub-critical zone is assured to be dominated by inverse bremstrahlung. A reflective dump operates at the critical surface - the fraction of the energy to be dumped is specified by the user.

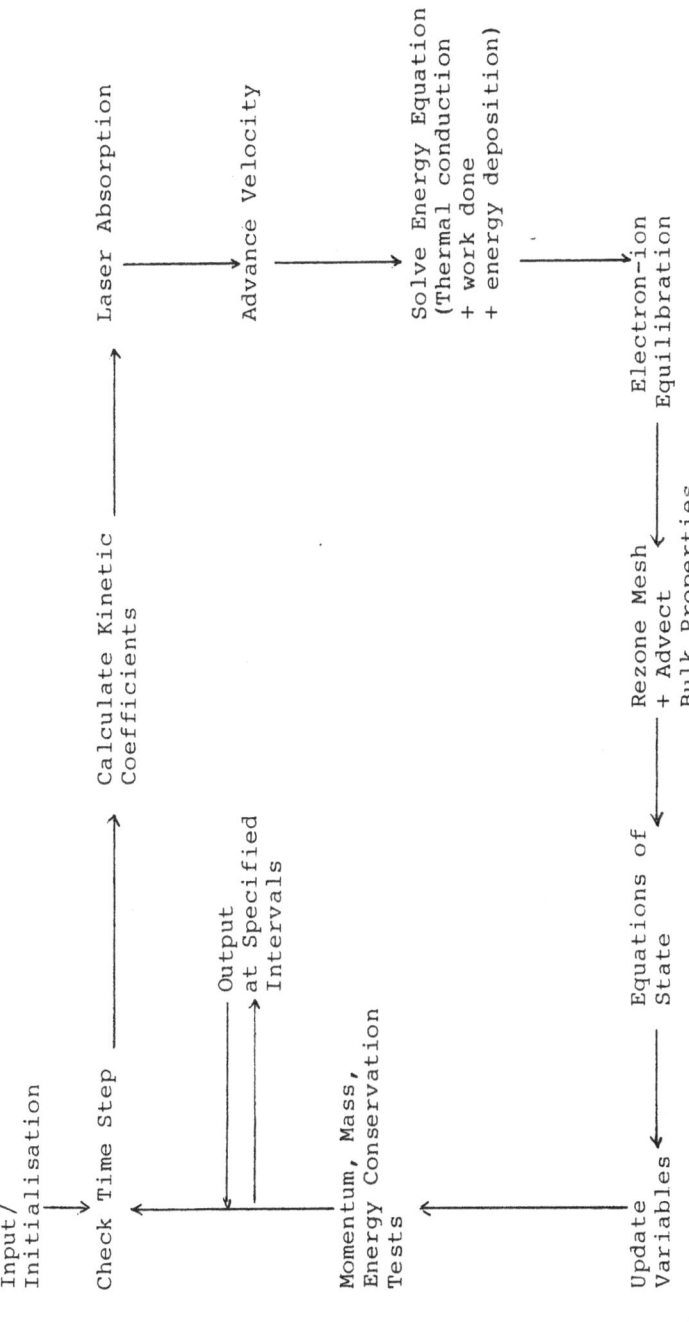

FIG. 2 THE STRUCTURE OF POLXM

213

3.4 Advancing the velocity (Sub:Accln)

The time dependent part of the fluid momentum equation is solved. The fluid velocity field is updated.

3.5 Energy Equation (Sub:Energy,ICC,Invert,Prod)

Thermal conduction is treated using a fully implicit five point method. The terms for the laser energy deposition and hydrodynamic work done are incorporated at this stage. The resulting five-diagonal matrix is solved using the well known Inverse Choleski Conjugate Gradient (ICCG) method. The matrix elements are evaluated in the subroutine Energy and ICCG is separated into three components: ICCG, INVERT and PROD.

3.6 Electron/Ion Equilibration (Sub:Equil)

Although the energy distributions of both electrons and ions are assumed to be Maxwellian, the two distributions are not necessarily in equilibrium. The exchange of energy between the two species is calculated using the classical equilibration rates.

3.7 Advection/Rezoning (Sub:Advect,Movez,Mover)

The advection of the bulk properties is strongly coupled to the rezoning of the mesh. Advect converts between thermodynamic variables and the properties transported by the advection subroutines Movez and Mover. A time splitting method has been adopted with each direction being treated separately. The rezoning may be uniform or quasi-lagrangian. The advection routines apply the FCT method to the Donor-cell and SHASTA algorithms. SHASTA is used in strong compression and Donor cell is used otherwise. FCT methods are of interest in many branches of fluid dynamics.

3.8 Equation of State

The equation of state is used to calculate thermodynamical properties. In the current version, the ideal gas equation is used for both electrons and ions, assuming constant ionisation. An unoptimised subroutine which evaluates the ionisation at each time step exists.

4. PARALLEL CONSTRUCTS

In order to minimise task creation and synchronisation overheads the decomposition into tasks should be as global as possible. In our experience it is relatively simple to find large scale parallelism in codes modelling physical phenomena, and to map it onto the architecture of the multiprocessor Cray X-MP supercomputers. The following examples illustrate the scope for exploiting high level parallelism.

4.1 Mesh Splitting.

The most obvious decomposition with CFD programs is the division of the computational mesh into sub-grids assigned to different tasks. This type of decomposition is very flexible with respect to the number of sub-grids and their arrangement. Fig 3a shows a decomposition where the grid is split about an axis to produce continuous sub-grids. Such decomposition is used in the advection routines of MPOLXM (the multitasked version of POLXM). Since POLXM uses a time splitting algorithm for the advection, the grid was split about the central Y-Y axis when calculating the advection in the Y direction and about the central X-X axis for evaluating the transport in the x direction. The decomposition shown in figure 3b is a subset of cyclic grid splitting where successive strips of the mesh are assigned to CPUs in a cyclic fashion.

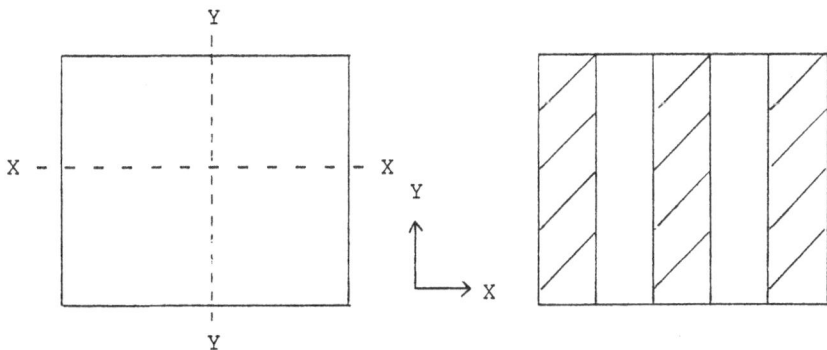

Figure 3a.   Figure 3b.

4.2 Fluid Species treated separately

With MPOLXM the energy equations for the ions and electrons are solved in parallel. Parallel processing for different fluid species is also possible for the ionisation calculations. When we have a plasma with several different elements, the ionisation calculations for different atomic species may be carried out in parallel. However, this may result in a load balancing problem (see section 5.2) if the nuclear charges of the atomic species vary significantly.

In another technique for multitasking ionisation calculations the ionic stages of a single species is divided into groups, each containing all levels between the stages with the highest and lowest nuclear charges within it. Three steps are involved in the ionisation calculations. The first is to evaluate the relative populations within the groups, in parallel. The relative populations of neighbouring ionisation stages belonging to different groups are calculated next. Finally the relative values are normalised to obtain the absolute populations.

4.3 Directional Parallelism

It is sometimes possible to update variables which have different directions associated with them in parallel e.g. the acceleration of the fluid in different directions may be evaluated in parallel. Similarly, the quasi-Lagrangian rezoning parameters and new mesh spacing in the two directions can be calculated concurrently.

4.4 Bulk Parameter Advection

We may evaluate the advection of different bulk fluid properties (e.g. mass, momentum, energy etc.) in parallel.

4.5 Algorithmic Parallelism

This is the class of large scale parallel constructs inherent to the algorithm used. An example is found in the Flux Corrected Transport (FCT) method for the advection equation. Generally, FCT requires two estimates for the advection - one from a high order scheme and the other from a low order scheme. Parallel processing of the two schemes may be used for multitasking the advection routine.

4.6 Parallel Hierarchies

Multitasking allows us to exploit parallelism at several levels simultaneously so that we may combine several different types of parallel constructs to form a parallel hierarchy which is exploited in practice e.g. grid splitting may be combined with one of the other constructs described above.

Where large scale parallelism is lacking we may resort to multitasking at the level of the innermost Do loops. It should be noted, however, that such limitations usually arise from the algorithm used and better performance my be obtained by changing the numerical method.

5. THE EFFICIENCY OF MULTITASKING

Buzbee [3] defines the efficiency, $E(P)$, of multiprocessing a single program using p processors as the ratio:

$$E(P) = \frac{\text{Speed-up } (P)}{P} \tag{1}$$

$$\text{Speed up } (P) = \frac{T_1}{T_2} \quad ,$$

where T_1 is the time required to execute the original program using 1 CPU, and T_2 is the execution time for the multitasked program, with P processors. The factors effecting $E(P)$ are discussed below.

5.1 Granularity/Multitasking Overhead

Initialisation and synchronisation of tasks, together with intertask communication, increase the number of CPU cycles required to execute a program. Clearly, the time for which tasks are active (i.e. their granularity) should be much larger than this overhead for efficient multitasking. It is important to note that the important parameter is the time for which the task is active and not the amount of arithmetic performed, i.e. For a given multitasking overhead, more arithmetic must be performed when executing in the vector mode, that in the serial mode, to achieve a specified multitasking efficiency.

5.2 Load Balancing

The computational load should be distributed as evenly as possible between the CPUs. This is the single most consideration when multiprocessing on a system with a few fast processors.

The following classification of task decomposition is particularly useful in discussing load balancing. Domain decomposition is where the computational domain is decomposed into sub-structures (e.g. grid splitting). With program decomposition we divide the program so that the tasks are responsible for executing different sets of calculations (e.g. The parallel calculation of the high and low order schemes for FCT).

With explicit numerical schemes it is easy to produce well-balanced tasks, with the usual restrictions imposed by data dependencies. The load balancing is deterministic for either type of decomposition (e.g. MPOLXM uses domain decomposition in the routines calculating the advection, and program decomposition in the routine calculating bulk properties used by them).

Load balancing is deterministic for program decomposition of a single implicit scheme, but may be relatively poor. With domain decomposition of implicit methods the load balancing may vary significantly with the parameter space examined.

In many cases, this may be circumvented by using dynamic load balancing where each task seeks for unfinished work at run time. Details of dynamic load balancing are given in ref. 4.

5.3 Vector Performance Degradation

Whenever possible we should avoid decompositions which reduce the vector lengths of the principal vectorised Do loops. The architecture of the Cray X-MP series is well suited to minimising the impact of reduced vector lengths - near asymptotic vector processing rates are achieved with relatively short vectors.

5.4 Shared Resource Conflicts

Depending on its size the central memory of the Cray X-MP
series is organised into 16, 32 or 64 banks. A bank is
reserved for 4 clock periods when referenced by a CPU.
Other CPUs are precluded from accessing the bank for this
period and tasks executing on them will be suspended if they
reference it. Hardware on the X-MP computers monitors
memory bank conflicts. Due to the large number of memory
banks on the multiprocessors, such bank conflicts do not
usually degrade the performance of multitasked code
markedly. If the performance is degraded significantly in
this way, the problem may be circumvented by minor changes
in the data structure of the code or by using a different
decomposition (e.g. cyclic grid splitting instead of central
splitting).

5.5 The Degeneracy and Dimensionality of Parallel Constructs

We define the degeneracy of a parallel construct to be the
number of processors onto which it maps without reducing
vector lengths. Similarly, we define the degeneracy of the
parallelism as the number of processors onto which it maps
without intertask synchronisation. The degeneracy of
parallelism inherent in physical problems tends to be small.
Hence, it is simpler to exploit a machine with a few fast
processors than one with a large number of slower
processors, having the same asymptotic speed.

When multitasking with the Cray X-MP series the following guide
line is generally valid. After satisfying the load balancing
criterion select the decomposition with the largest degeneracy
and dimensionality.

6. PERFORMANCE

Software available allows the development and execution of
multitasking code on the Cray X-MP/1 which has only 1 CPU.
MPOLXM was developed on a single processor. The simulated
performance for a two processor system predicts a speed up
of 1.67, implying a multitasking efficiency of 0.8.

Analysis of the execution time shows that the speed-up ratio
(T_1/T_2) has the following components;

$$\frac{T_1}{T_2} = \boxed{\begin{array}{c} 0.03 \\ \hline \text{overhead} \end{array}} + \boxed{\begin{array}{c} \frac{0.5(10.2)}{(16)} \\ \hline \text{ICCG} \end{array}} + \boxed{\begin{array}{c} 0.24 \\ \hline \text{other code} \end{array}} \cdot \quad (2)$$

The overhead introduced by multitasking routines is about
3%. The principal cause of the additional reduction in the
multitasking efficiency is the fact that the time required

218

to solve matrices from the implicit treatment of energy equations differs significantly for the ions and electrons. Hence, the load is not well balanced for the part of the code in which the ion and electron energy equations are solved in parallel. In the equation above ICCG is the ratio of the total time spent in the matrix solvers by the multitasked and original codes. This illustrates the point regarding load balancing when domain decomposition is used with an implicit scheme.

Table 1 shows the Cray X-MP performance for other multitasked programs.

Table 1: Speeds of multitasked programs normalised
to the performance using 1 CPU of a Cray X-MP
Supercomputer.

Code	Cray X-MP/2	Cray X-MP /4
Particle in Cell Plasma Physics Code	1.90	3.48
Weather Prediction Model	1.91	3.50
Monte Carlo Model for Radiation Transport	1.96	3.75

7. CONCLUSIONS

Substantial progress with CPU-bound problems is possible by exploiting the multitasking features of the Cray X-MP multiprocessors. Each CPU of Cray X-MP supercomputers is an enhanced version of the central processor of the Cray-1. Multitasking is an additional dimension of parallel processing on these machines which compliments the vector processing capabilities of each CPU.

Whilst vectorisation operates at the level of the innermost Do loops, in FORTRAN, the division of the job into sub-tasks should be as global as possible. With the current generation of Cray supercomputers multitasking can speed up the execution by a factor ≤ 4. In our experience it is relatively simple to find large scale parallelism in simulation codes. The software available simplifies efficient exploitation of such parallelism.

In this paper we have examined the factors important in efficient multitasking. With the Cray X-MP series the single most important consideration is load balancing.

When we have a choice between well balanced parallel constructs, the one with the highest degeneracy and dimensionality should be exploited in order to minimise communication overheads and maximise vector lengths.

Load balancing is deterministic for domain decomposition with explicit schemes - and program decomposition of both explicit, and implicit, schemes. If domain decomposition is used with an implicit method the load balancing may vary significantly with the parameter space examined. Dynamic load balancing is an interesting technique for this combination.

To summarise, Multitasking with the Cray X-MP series offers a significant reduction in the elapsed times for CPU bound problems. Moreover, experience with these systems will prove invaluable in the future as designers look to progressively more parallel computer architectures.

References:

[1]   R.W. Hockney and C.R. Jesshope
       Parallel Computers (Adam Hilger Ltd.) 1981.

[2]   G.J. Pert, J. Compt. Phys, vol 49,P1,1983
       Rutherford Appleton Laboratories
       Annual Report of the Laser Facility, 1982, 1984.

[3]   B. Buzbee Proceedings of the Vector and Parallel
       Processing Conference (Oxford), 1984 - to be
       published.

[4]   Multitasking Users Guide - published by Cray
       Research Inc. (Sn.0022), 1440 Northland Drive,
       Mendota Heights MN55120.

Development of an Atmospheric Mesoscale Model on a CRAY - Expericences with Vectorization and Input/Output

Hans Volkert and Ulrich Schumann

DFVLR, Institut für Physik der Atmosphäre
D-8031 Oberpfaffenhofen, Fed. Rep. of Germany

Summary

Some concepts and experiences are discussed from the present development of a three-dimensional model for atmospheric flows in the mesoscale (typical lengthscale $L \leq 250$ km) on the CRAY-1/S computer of the DFVLR. After an introduction to the range of physical problems and adequate numerical schemes to tackle them two more technical aspects are dealt with. First, an integration algorithm, which minimizes input/output operations, is introduced together with figures that show the capabilities of different software and hardware components for input/output. Then, the emphasis goes to the pressure solution as an important subtask, the vectorization capabilities of existing software and gains due to its restructuration. Calculations involving the entire code (dealing with the Taylor-Green vortex in a 64×36×64 grid) and a discussion of the technical aspects' impact on three dimensional Navier-Stokes codes conclude the paper.

Introduction

During recent years mesoscale modelling constitutes a continuously growing field of meteorology (see [1] for general reference and an extensive bibliography). Based on experiences with imported codes (see [2,3,4]) the design and implementation of MESOSCOP (Mesoscale Flow and Convection Model Oberpfaffenhofen) was initiated at the Institut für Physik der Atmosphäre. These existing models suggest the necessity of a flexible representation of physical processes, higher spatial resolution and computational speed and, perhaps most important, *one* basic concept of data management for a variety of applications. Among the topics of interest to the meteorological community are mountain-valley wind systems, airflow over mountains (e.g. Föhn events), orographically induced precipitation, turbulence in boundary layers and clouds, thunderstorms (possibly with hail) and dispersion of pollutants.

For such calculations a three-dimensional (3-d), non-hydrostatic model is prerequisite. MESOSCOP is based on the budget equations for density (ρ), momentum (ρV) and several scalars ($\rho \psi_k$, k=1,..,K; e.g. entropy and water phase concentrations):

221

$$\frac{\partial(\rho \Psi_k)}{\partial t} + \text{div}(\rho v \Psi_k) = -\text{div}(F_k) + Q_k$$

$$F_i \frac{\partial \rho}{\partial t} + \text{div}(\rho v) = 0 \qquad\qquad F_i \in \{0,1\} \qquad\qquad (1)$$

$$\frac{\partial(\rho v)}{\partial t} + \text{div}(\rho v v) = -\text{grad}(p) - \rho g + \text{div}(F),$$

where p stands for pressure, **g** for the constant gravity vector, **F** for the stress tensor, f_k and Q_k for the friction vector and for source/sink terms in the scalar budgets, respectively. The budget equations are supplementd by equations of state relating density, pressure, temperature (T) and the other scalars

$$\rho = Z_\rho(p, \Psi)$$

$$T = Z_T(p, \rho, \Psi). \qquad\qquad (2)$$

As the airflows of interest are controlled by advection an explicit treatment is possible for most terms in equations (1). Only terms describing fast processes (e.g. sound waves or phase changes) are treated implicitly to avoid strong limitations for the appropriate timestep. The implicit treatment of the pressure gradient term and the divergence of momentum in equations (1) lead to an elliptic equation for the pressure increment (Δp) per timestep:

$$\text{div}(\text{grad}(\Delta p)) - F_i \subset \Delta p = F. \qquad\qquad (3)$$

Depending on the filter parameter (f_i) switched on (anelastic approximation) or off (full incompressibility) this equation is of Poisson or Helmholtz type (with Helmholtz parameter c), respectively.

The model equations are discretized via finite differences on a staggered grid (see [5] for details). According to the two main areas of application two versions of MESOSCOP exist, one in Cartesian coordinates for studies of deep convection including cloud physics and the other in terrain following coordinates for dry airflow over irregular terrain [5,6]. Common to both is the management of automatic data transfer between core and disk memory for grid sizes too large to fit in the core all at once. A second mutuality lies in the inversion of the elliptic pressure equation (3). In the next sections details of the data management and the vectorizing capabilities of fast Helmholtz solvers are discussed and results of Taylor-Green vortex calculations are presented.

Input/output considerations

The introduction of a 3-d grid as approximation to the continuous computational domain implies that a number NF of 3-d fields has to be stored in the computer, each containing (IM+2)*(JM+2)*(KM+2) locations (IM, JM, KM within the computational domain in x-, y- and z-direction, respectively, and 2 for the boundary values on either side for every direction). Taking as a low estimate NF=30 (the exact figure depends on the number of scalars taken into account) and IM=JM=KM=30 we realize that approx. 10^6 words of core memory would be necessary without data transfer from core to disk. As the CRAY-1/S at the DFVLR offers only 855000 words minus storage for the compiled code it is essential to provide a means for data transfer if the model should not be restricted to quite limited (or only 2-d) domains.

The dynamic core (DC) subroutine package (a preliminary version is described in [7]) provides such a means. Any 3-d field is subdivided into x/z planes (see Figure 1), so-called blocks. Each block of length IBLKCC provides space for the field's values within the specific plane plus two control words for data checks (IBLKCC=(IM+2)*(KM+2)+2). All blocks are stored in a buffer vector ACC of length ILBUCC (e.g. ILBUCC=500000) and uniquely identified by their slice number J, field number IFCC and start address JFJ of the first element (at i=k=1; small cube in Figure 1) within ACC. There are three main tasks which manipulate ACC:

i)   to provide space for reading data (routine DCR),
ii) to provide space for writing newly created data to (routine DCR) and
iii) to provide space for updating data (e.g. reading, manipulating and writing; routine DCU).

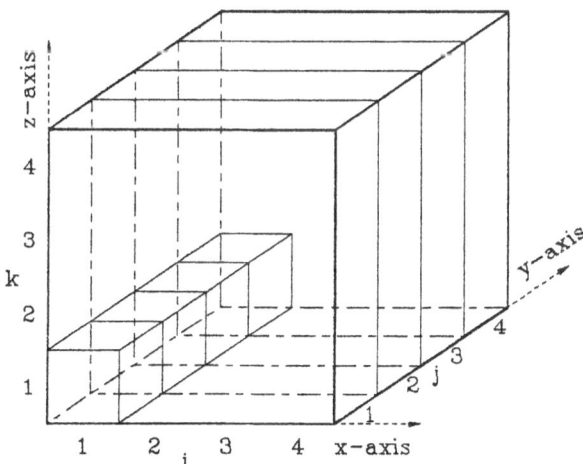

Figure 1.   Schematic sketch of 3-d domain (IM=JM=KM=2) with boundary cells, divided in x/z planes. Small cubes at i=k=1 indicate the position of field elements referenced by starting addresses.

223

```
                    DIMENSION ACC(ILBUCC)
                    COMMON ...
                    DO 10   J=1,JM+1
                    CALL DCR      ( J ,IRHCC ,JRH)
   level            CALL DCR      (J+1,IRHCC ,JRHP)
   of               CALL DCR      ( J ,IVCC  ,JV)
   data             CALL DCW      ( J ,IRHVCC,JRHV)
   management       CALL RHVJ     (ACC(JRH),ACC(JRHP),ACC(JV),ACC(JRHV),J)
                    CALL DCDELE ( J ,IRHCC)
                    CALL DCDELE (J+1,IRHCC)
                 10 CALL DCDELE ( J ,IVCC)
```

```
                    SUBROUTINE RHVJ (RH,RHP,V,RHV,J)
                    DIMENSION  RH(*),RHP(*),V(*),RHV(*)
   level            COMMON ...
   of               DO 10   I=1,(IM+2)*(KM+2)
   physical      10 RHV(I) = V(I) * (RH(I)+RHP(I))*0.5
   formulae         RETURN
                    END
```

Figure 2. Example for levels of data management and physical formulae.
           After reading in appropriate planes of ρ and v and preparing space
           for the results the momentum component ρv is evaluated within a
           staggered grid. Communication between subroutines via parameter
           list and common blocks (not given here). The second parameter in
           each DC-routine call is the specified field number (generally
           called IFCC).

Further tasks comprise the marking of blocks as presently not necessary in
main memory (routine DCFREE) or as never more necessary (routine DCDELE).

If a request to ACC exceeds its length ILBUCC, blocks marked as 'presently
not necessary' are transferred to disk in order to make space available.
After a read or update request the system checks whether the block is already
in core. If not, it is read from disk. Routine DCR leaves a copy of the read
block on disk while use of routine DCU implies the block's deletion on disk
and its later re-transfer when its contents has been changed. The bookkeeping
is made by tables for the blocks on disk and in core, respectively. A facility for randomly accessing the file on disk is essential. Examples, how that
is realized on a CRAY, will be given below. The quantity of data transferred
at a time is one block (IBLKCC words).

Figure 2 exemplifies the data management's impact on the general structure of
the code. The level of physical formulae (here: calculation of momentum component ρv within a staggered grid) is roofed by a level of data management,
where the appropriate blocks are made ready, the physical routine is called
using the provided starting addresses within ACC and unnecessary blocks are
freed or deleted. In the given example the original loops over indices I and
K are united to one long loop. If necessary, explicit reference to 2-d arrays
is just as well possible.

Step	Plane-relation in staggered grid	Comment
P2	$U_j = F1(U_j', P_j)$ $V_j = F2(V_j', P_j, P_{j+1})$ $K_j = F3(U_{j-1}, U_j, U_{j+1}, V_{j-1}, V_j)$	U from U' and $\partial P/\partial x$ V from V' and $\partial P/\partial y$ diffusion coeff. from U, V
P1	$U_j' = F4(U_{j-1}, U_j, U_{j+1}, V_{j-1}, V_j, K_{j-1}, K_j, K_{j+1})$ $V_j' = F5(U_j, U_{j+1}, V_{j-1}, V_j, V_{j+1}, K_j, K_{j+1})$ $P_j = F6(U_j', V_{j-1}', V_j')$	U' from explicit terms V' " " " source term for press. equ.
D	$P = F7(P)$	solution of press. equation

Figure 3. Simplified example for integration structure. Natural order of prognostic (P1, P2) and diagnostic (D) parts: P1, D, P2. Best order within integration (P2, P1, D) allows one single loop over index j. Details in Figure 4.

During integrations on large grids (typically more than 30^3 grid points) input/output (i/o) operations cannot be avoided. On a CRAY i/o to and from ordinary disks is significantly slower than the calculations themselves and in synchronous mode it prevents calculations until it has finished. As we presently use synchronous i/o, we aim at minimizing the number of i/o operations. From Figure 1 it is obvious that difficulties in that respect arise from neighbourhood relations along the y-axis (e.g. by $\partial/\partial y$-terms) as different planes become involved.

Figure 3 gives a simplified example, which highlights the structure of the problem. Suppose all variables are ready at some time level. In a first prognostic step P1 preliminary velocity values U_j', V_j' are calculated from the terms treated explicitly in the momentum budget (here the difference between

Field name	Program name	Plane# in domain (j) 2 3 4 5 6	Technical indices jsub jadd	Example jj
P				
U'				
U	F1		0 0	6
V'				
V	F2		1 0	5
K	F3		2 0	5
V'	F5		3 1	4
U'	F4		4 2	4
P	F6		4 2	4

Figure 4. Simplified example for 'back slide' structure in the i/o minimized integration procedure without 'post processing'. o: read field; +: calculate and write field; example for j=6 illustrates the relation jj=j-jsub+jadd; see text for details.

225

site / machine	DFVLR / CRAY-1/S					CRI / CRAY X-MP 48			
i/o method	FT-77	READDR/WRITDR							
COS version	1.13			1.14(C)		1.14	X.15	1.14	X.15
Device type	DD19					DD49		SSD	
access time (s)	-	517	523	590	449	1455	994	11	34
blocked time (s)	537	334	306	283	414	951	541	<.005	<.005
i/o rate (Mw/s)	.079	.127	.138	.149	.102	.044	.078	>8.46	>8.46

Figure 5.   Input/output diagnostics for testproblem (4 timesteps, 64^3 mesh, transfer of 42.3 Mwords) in dependence of machine type, i/o method, operating system version and external storage device. The paradox that DD49-times are longer than those with DD19 is explained in the text.

velocity and momentum is neglegted), and from these the source term of the elliptic pressure equation (see [5] for a more detailed derivation). The j-dependence due to derivatives and the staggered grid is given explicitly. A diagnostic step D yields the pressure increment by the inversion of the elliptic operator div**grad**. In order to make use of fast direct solvers the 3-d pressure field has to be treated as one entity in core. In a second prognostic part P2 the preliminary velocities are updated to the final ones by taking into account the just obtained, continuity consistent pressure, and diffusion coefficients (first order closure) are calculated from these.

Within an integration run this order can be rearranged to P2, P1, D if a preparation and postprocessing step are provided for the begin and the end of the integration. Now only one single j-loop over P2 and P1 is necessary if the 'back slide procedure' sketched in Figure 4 is followed. The path starts at j=2 with reading P_2 and U_1' and then calculating and storing U_2. As P_3 is necessary for evaluating V_2 the path stops, starts again at j=3, the procedure is repeated for U_3 and after a slide back to j=2 V_1' is read and V_2 calculated. U_4 (possibly under periodic boundary conditions) being not available we jump to j=4, continue a parallel path as before to V_3 and end up with K_3. Proceeding in the same manner we finally reach P_4 having started at j=6. In a postprocessing step all missing planes ($K_2,\ldots,P_6$) are evaluated.

Significant for the structure of the diagram are the technical indices jsub (after how many restarts a variable is reached for the first time) and jadd (how many planes stay empty till postprocessing). The final j-loop looks like j=2,JM+1+max{jsub}, whereas the planes of the different fields are addressed by jj=j-jsub+jadd as exemplified in Figure 4 for j=6.

The back slide diagram of the entire code is, of course, much longer, but it exhibits the same structure with two back slides (max{jsub-jadd}=2). This ensures that during one timestep every plane is read and written only once (except, perhaps, at the boundaries). Comparisons with a similar code, which contains several straightforeward loops over index J, lead to the estimate of a i/o reduction factor of five due to the back slide algorithm.

To obtain i/o diagnostics for the scheme just introduced a test problem (Taylor-Green vortex [see below], IM=JM=KM=64, 4 timesteps, transfer of 42.3 Mwords in portions of 4608 words) is run on the CRAY-1/S of DFVLR and the CRAY X-MP/48 of Cray Research Inc. (CRI) using two software methods (standard FORTRAN 77 direct access read/write and the Cray library routines READDR/WRITDR), three device types (disks DD19 and DD49 as well as the solid state storage device SSD with two high speed channels) and three versions of the Cray operating system. In Figure 5 the total access time (from file $STATS, only available for READDR/WRITDR), the time blocked for i/o (given in the logfile if OPTION,STAT=ON is invoked) and estimated i/o rates (42.3 Mwords / blocked time) are compiled.

The tests with the configuration at DFVLR reveal: the standard FORTRAN i/o version is considerably slower than the direct access routines from the Cray library; the results depend on the job mixture in the machine and, thus, are hardly reproducible; the total access time might be reduced by the new feature of assigning contiguous disk space [COS 1.14 (C)]; the estimated i/o rates amount at the best to 30% of the maximum figure for disk type DD19 (0.5 Mwords/s). Recent runs with the method applied by Kordulla (BUFFER IN/OUT combined with routine SETPOS; see [8]) yield similar i/o times as the READDR/WRITDR case.

The comparisons carried through at CRI show: transfer on disk type DD49 appears to be approx. twice as slow compared to DD19, although the maximum figure lies 2.5 times higher (1.25 Mwords/s); the explanation for that lies, as we hear, in the use of a test environment, which combines many types of i/o devices, and in a bug (meanwhile fixed) within the input/output subsystem (IOS); use of the SSD reduces the total access time by approx. two orders of magnitude, while the diagnostic for blocked time can be only roughly assessed (just two decimal places [.00] are given); the estimated i/o rate of exceeding 8.5 Mwords/s for an installation with two high speed channels fits to the figure of 5 Mwords/s published by CRI for the transfer of similar sized data segments over one such channel; as the transfer rates for only a few sectors (each of 512 words) become neglegible compared to the access time, new system software is necessary for efficiently organizing i/o to and from a SSD; this is presently developped by CRI under the label 'asynchronous queued i/o (AQIO)'.

Experiences with vectorization

Modern vector computer achieve their great speed particularly from the special treatment of innermost do-loops (vectorization), provided certain rules are obeyed within these (compare [9]). While the main body of MESOSCOP is designed with the capabilities and peculiarities of a CRAY vector computer in mind, the inversion of the elliptic pressure equation (3) relies on existing software. In this section experiences with its vectorization are reported.

	Vector mode	total time	L	H	POISSX/ POISSV	pre-/post processing	FFT + FFT $^{-1}$
POISSX	OFF	2.73	0.93	1.65 {1}	1.25	0.22	0.18
	ON	1.37 (1.34)	0.20	1.13 (1.10)	1.00	0.08	0.05 CAL:(0.02)
POISSV	OFF	2.57	0.93	1.56 {2}	1.06	0.22	0.18
	ON	0.63 (0.60)	0.20	0.38 (0.35)	0.25	0.08	0.05 CAL:(0.02)

Figure 6. Computing times (in s) on a CRAY-1/S for the different components in equation (4) using a 62×30×30 domain (one iteration). {1}: 2435 words workspace; {2}: 7535 words workspace.

The discrete form of the pressure equation implies the inversion of a 7-point difference operator H for a Cartesian coordinate system (variable meshsizes are permitted in the vertical) and of a much more complicated 25-point operator L for terrain following coordinates. H is directly invertible, whereas the inversion of L can be achieved by a block iteration according to

$$H(\Delta p_{\mu+1} - \Delta p_{\mu}) = f - L(\Delta p_{\mu}) \qquad \mu = 0, 1, \ldots \qquad (4)$$

with iteration index μ, pressure increment Δp and source term f. A detailed derivation of H and L, the definition of a test problem with homogeneous Neumann-type boundary conditions and convergence rates for the block iteration are given in [5]; here we restrict the discussion to computing times due to vectorization for the different components in one iteration step in equation (4). We note that an alternative solution method lies in the application of multigrid techniques (see [10]).

Subroutine H3DNDC inverts H by applying in sequence a fast Fourier transformation (FFT) in y-direction (combined with pre-/postprocessing due to the Neumann boundaries; see [11]), the independent solution of JM 2-d problems (routine POISSX; see [12]) and the return to physical space (FFT $^{-1}$). Due to the recursive structure of POISSX the vectorization gain is quite limited, while a newly constructed version POISSV solves all JM 2-d systems simultaneously at the expense of more work storage (see [13]). FFT and FFT $^{-1}$ are realized by routine FFT99 (by C. Temperton; available in FORTRAN or in Cray Assembler Language [CAL]).

Figure 6 contains computing times for a test problem of IM=62, JM=30, KM=30 grid points in x-, y- and z-direction, respectively, with the Cray vector mode both turned on and off. Aside of the absolute figures the key result lies in the fact that a factor of 4.5 time gain due to vectorization is only achieved after reconstruction of an essential subroutine. Otherwise the speed up factor equals 2.

228

Calculation of the Taylor-Green vortex

It is advisable to test a newly assembled code against a simple analytical solution first. Taylor [14] gives a special 2-d solution of the Navier-Stokes equations under the assumption of constant density and constant kinematic viscosity ν in the friction term:

$$u(x,y,z,t) = -U k_z \cos(k_x x) \sin(k_z z) \exp(-[k_x^2+k_z^2]\nu t)$$

$$w(x,y,z,t) = U k_x \sin(k_x x) \cos(k_z z) \exp(-[k_x^2+k_z^2]\nu t) \qquad (5)$$

$$v(x,y,z,t) = 0 .$$

U denotes a velocity amplitude and k_x and k_z wavenumbers for the respective directions. The streamline configuration connected to equations (5) is known as the Taylor-Green vortex. To test the integration scheme for positive definite scalars [15] we consistently introduce a temperature T' with amplitude T:

$$T'(x,y,z,t) = 1 + T \cos(k_x x) \cos(k_z z) \exp(-[k_x^2+k_z^2]\nu t) . \qquad (6)$$

The analytical expression for the kinetic energy integrated over the entire domain V reads:

$$E_{kin} = \frac{1}{2} \iiint_V d\tau (u^2+v^2+w^2) = \frac{1}{8}(k_x^2+k_z^2) U^2 \exp(-2[k_x^2+k_z^2]\nu t) . \qquad (7)$$

Taking the constants as $\nu = 0.01$ m^2/s, $k_x = 2\pi$ m$^1 = k_z$, $U = 1/(2\pi)$ m/s, $T = 1$ K we integrate the Cartesian version of MESOSCOP from the initial conditions of equations (5,6) at t=0 for a unit cube domain ($-.25$ m $\leq$ x,y,z $\leq$ .75 m) with periodic boundary conditions in all three directions. Figure 7 displays results from two runs with different spatial resolution. The calculated exponential decay for velocity and temperature maxima and for kinetic energy (straight lines when ordinate is scaled logarithmically) lies about 5% under the theoretical value when 8 discrete points come to one wavelength. Increasing the spatial resolution by a factor of 8 lowers the relative deviation between calculation and theory by a factor of 70 to about 0.07%. This agrees very well with what is expected from a differencing scheme of second order accuracy. The computing time amounts to 8.6 s/timestep or 52 μs/(timestep×gridpoint) for the large case.

Conclusions

The final aim of a numerical modelling project such as MESOSCOP is that atmospheric physicists want to get more insight into the detailed three-dimensional nature of atmospheric flows in scales smaller than these dealt with by the national weather services. A number of concrete applications

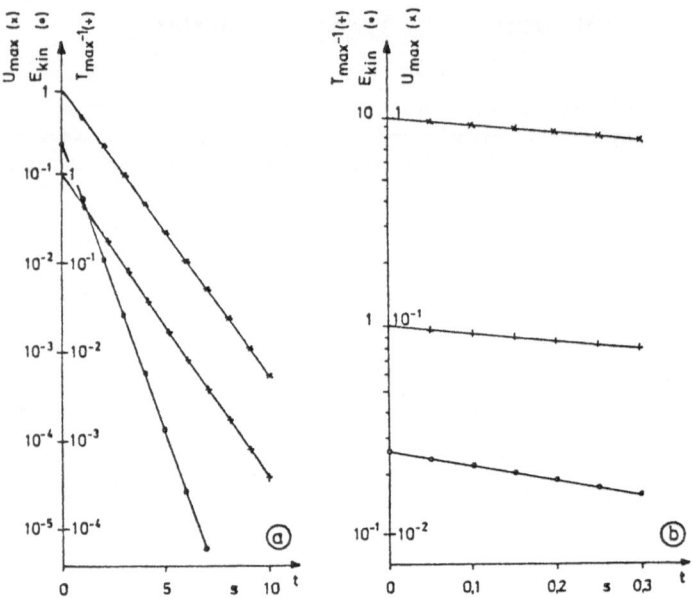

Figure 7. Calculated exponential decay of velocity (U_{max}) and temperature ($T_{max}-1$) maxima and of the kinetic energy (E_{kin}).
Part a: 8×8×8 domain, ν = .01 m^2/s, Δt = .02 s, 0.41 CPU s / Δt.
Part b: 64×36×64 domain, ν = .01 m^2/s, Δt = .002 s, 8.6 CPU s / Δt.

waits to be studied with the tool of a reliable numerical model. Besides physical soundness and problem dependent flexibility effectiveness in computing is essential.

Data segmentation is inevitable for any realistic 3-d flow problem with the presently available main frame core size. Organising the corresponding input/output synchronously over one channel to a random access file makes the code flexible (e.g. with respect to boundary conditions). A special integration algorithm ensures that i/o operations are minimized (one read and one write per block and timestep). Nevertheless, it has to be admitted, that alternative methods allowing more i/o operations, but using several sequential files and the asynchronous mode widen the i/o bottleneck for special problems.

Hardware components, which drastically cut down the total access time for input/output without the necessity of changing the code structure, are available. For Cray installations these are the input/output subsystem (IOS), which makes possible the distribution of one file onto several disks (file striping), and - most powerful - the solid state storage device (SSD). But these features are expensive (at least from our perspective) and presently not at our disposal.

Realising the fast development on the hardware and system software side and anticipating significant upgrading of the facilities available to us we presently put first priority to the physical side of numerical modelling. This includes tests through a hierarchy of more and more complex physical problems (the Taylor-Green vortex constitutes the first level) and the completion of the version in terrain following coordinates [16].

The vectorization of the innermost do-loops is mostly straightforward, once the programmer knows about this feature from the very beginning. New compiler releases ease application in that respect (e.g. directive SHORTLOOP and option KILLTEMP of CFT 1.14 are found useful). The 'recursivity problem' in the diagnostic part (using 'pre-vector' software) is solved by the investment of work storage.

Finally, we state the banal experience that the way through changing system software components (compiler, operating system), sometimes not well documented diagnostics and general information from the manufacturer is quite arduous for the end-user of advanced computer systems. It would be appreciated, if the material presented here helps to intensify the contacts and exchanges between users from the physics' side of problems and the computer companies' application analysts.

Acknowledgement

The kind assistance of H. Cornelius (Cray Research GmbH, München) for carrying out test runs at Cray Reasearch Inc. in Minneapolis, U.S.A., is gratefully acknowledged. Furthermore, he provided helpful advice and guidance on our way through technical details. G. Körner is thanked for preparing Figure 7.

References

[1] PIELKE, R.A.: "Mesoscale Meterological Modeling", Academic Press, Orlando (1984).

[2] SOMIESKI, F.: "Numerical Simulation of Mesoscale Flow over Ramp-Shape Orography with Application of a Special Lateral Boundary Scheme", Mon. Wea. Rev. 112 (1984), pp. 2293-2302.

[3] HOINKA, K.-P.: "A comparison of numerical simulations of hydrostatic flow over mountains with observations", Mon. Wea. Rev. 113 (1985), in press.

[4] WENDLING, P.: "A three dimensional mesoscale simulation of topographically forced rainfall on the northern side of the Alps", Ann. Meteorol. (N.F.) Nr. 19 (1982), p. 27.

[5] SCHUMANN, U., VOLKERT, H.: "Three-dimensional mass- and momentum-consistent Helmholtz-equation in terrain-following coordinates", in: W.Hackbusch (Ed.), "Efficient Solutions of Elliptic Systems", Vieweg Series "Notes on Numer. Fluid Mech." Vol. 10, Braunschweig (1984), pp. 109-131.

[6] CLARK, T.L.: "A small-scale dynamic model using a terrain-following coordinate transformation", J. Comput. Phys. **24** (1977), pp. 186-214.

[7] SCHUMANN, U.: "Dynamische Datenblock-Verwaltung in FORTRAN", KFK External report 8/74-2, Karlsruhe (1974).

[8] KORDULLA, W.: "On the efficient use of large data bases in the numerical solution of the Navier-Stokes equations on a CRAY computer", this volume (1985).

[9] GENTZSCH, W.: "Summary of the workshop", this volume (1985).

[10] THOLE, C.-A.: "Multigrid for anisotropic operators in three dimensions", contribution to the Workshop "INRIA - GMD", Versailles, January 21-23, 1985.

[11] WILHELMSON, R. and ERICKSEN, J.: "Direct solutions for Poisson's equation in three dimensions", J. Comp. Phys. **25** (1977), pp. 319-331.

[12] SCHUMANN, U. and SWEET, R.: "A direct method for the solution of Poisson's equation with Neumann boundary conditions on a staggered grid of arbitrary size", J. Comp. Phys. **20** (1976), pp. 171-182.

[13] SCHMIDT, H., SCHUMANN, U., ULRICH, W., VOLKERT, H.: "Three-dimensional direct and vectorized elliptic solvers for various boundary conditions", DFVLR-Mitteilung 84/15, (1984).

[14] BATCHELOR, G., (Ed.): "Scientific papers of Sir G. I. TAYLOR", Vol. II, Cambridge University Press (1960), Paper 16: The decay of eddies in a fluid (written in 1923).

[15] SMOLARKIEWICZ, P.: "A fully multidimensional positive definite advection transport algorithm with small implicit diffusion", J. Comp. Phys. **54** (1984), pp. 325-362.

[16] VOLKERT, H. and SCHUMANN, U.: "Development of an atmospheric mesoscale model - first results of the version in terrain following coordinates", Proc. Sixth GAMM Conf. on Num. Meth. in Fluid Mech., Göttingen, 25-27 Sept. 1985, Vieweg Series "Notes on Numer. Fluid Mech.", in preparation.

TECHNIQUES FOR SPEEDING UP CLIMATE MODELS ON THE

CYBER 205

Mavis K Hinds

Meteorological Office, London Road, Bracknell, Berkshire RG12 2SZ, England

SUMMARY

A number of different methods are described of reducing the elapsed time taken to run a climate model on a Cyber 205 computer whilst improving the understanding of the meteorological processes and the relationship between the model and the real atmosphere.

The UK Meteorological Office has had a Cyber 205 computer for almost four years with one megaword of memory, two vector pipelines, six 919/2 disks (total 450 megawords) and an RHF link to an IBM front-end. In February 1985 the memory was increased to two megawords. It has a number of purposes, but the two primary ones are operational real-time weather forecasting and research into climate modelling. The author has been responsible for programme organisation for climate modelling, and this paper will concentrate on methods of reducing elapsed time in the computer.

The climate model simulates the global atmosphere and information is stored for points on a regular latitude/longitude grid. The model is written so that the number of grid points in the three directions, east/west, north/south, up/down can be set at run-time using PARAMETER statements. An early decision had been needed regarding the arrangement of the data in the three-dimensional array (see Figure 1). On scalar computers it had been found to be efficient to store all information for one column of data together, and the model had also run satisfactorily on the CRAY-1 in this manner because of the constant stride between vector elements. However, on the Cyber 205 it was necessary for vector elements to be contiguous in memory, so that storage in vertical columns would have been very inefficient. Consideration was given to two alternative methods of storing the data in memory, either storing all the information for one vertical slice or slab together, or storing all the information for one horizontal field together (see (b) and (c) on Figure 1). The former would lead to a vector length of the order of 1000 and the latter of the order of 10000.

Conservation is very important in climate modelling and therefore the flux form of the dynamical equations is used. For this reason, at any one time during the calculations information is needed from only two adjacent "grid-boxes" in a north/south or east/west direction, but from all the "grid-boxes" in the up/down direction (because of the hydrostatic approximation). The advantages and disadvantages of the two alternative methods of storage are as follows:-

Storage by vertical slices or slabs	Storage by horizontal fields
(i) medium length vectors (of order 1000) .	(i) long vectors (of order 10000)
(ii) latitude constants can be used as scalars, enabling a speeding up by use of linked triads	(ii) fields of latitude constants are needed, leading to increased use of working-space, extra programme time to produce them, and slower computation due to lack of linked triads.
(iii) there is adequate working-space available	(iii) very large working-space is needed because it is proportional to the vector length
(iv) paging is possible because as soon as the calculation for any vertical slab is completed, that part of memory can be paged out.	(iv) no paging is possible, because the information at other horizontal levels must always be available.

Since the start-up time in vector arithmetic is the equivalent to 100 or 200 calculations (for 64 or 32 bit arithmetic respectively), the effect of vector length favours the arrangement in fields, but the difference is only 10 or 20% of the total computing time. The other differences all point to it being advantageous for the information to be arranged in vertical slabs. Since the use of horizontal fields would have required an impossibly large amount of working-space this led to a decision in favour of data storage in vertical slabs.

With only one megaword of memory (ie 16 large pages) there were versions of the model which required more memory than was available to the user (ie 15 large pages rather than 14). We discovered that a noticeable amount of elapsed time (about 30 seconds per model-day) was being wasted on paging during the dynamical routines. The programme sweeps through the memory in a cyclic fashion, and data is mapped on to large pages to reduce the paging. However, the computer will always page out the large page of memory accessed least recently, and with a cyclic programme this is the page which is needed next. Thus for this type of programme the automatic paging system is very inefficient. However, using the addresses in descriptors it is simple to compute where the large page boundaries are in the grid. At the end of processing a row containing a large page boundary Q5ADVISE can be used to page out the whole large page of memory which is no longer needed, and a further Q5ADVISE can be issued to page in the next large page of memory needed. These two transfers will be chained together and computing can continue in parallel with the transfers. We were thus able to eliminate almost completely the 30 seconds wasted time per model day.

The physics routines (ie the boundary-layer, convection, rain and radiation schemes) work on each column of data independantly and therefore for this part of the calculation the data is block-gathered into fields or parts of fields tailored to the working-space available, and then block-scattered back into position afterwards. The block-gather and block-scatter instruct-ions (Q8VXTOV and Q8VTOVX with G-bit 6 = 1) are very fast on the Cyber 205 and can be used to re-order matrices of data very efficiently.

The basic model programmes were written in 1980 when there was no 32-bit facility in the compiler, therefore for convenience they were written using 64-bit arthmetic even though for most purposes this accuracy is unnecessary. The WHERE facility was not available either and therefore in order to use the bit-control which was necessary in the physics routines, Q8calls were used. Once the programme had been developed on a coarse mesh grid, a finer mesh was introduced and this immediately led to problems of memory size and paging time. Most of the data was therefore reduced to 32-bit format and the routines dealing with the dynamical equations were transliterated into 32-bits Q8calls. Some of the physics routines have similarly been changed to 32-bit arithmetic, but for the others the word-length of the data is conveniently changed at the same time as the data is block-gathered and scattered.

The same COMMON area is used for work-space by all the different routines in the model and this has been mapped on to a large page boundary to cut down paging. For the physics routines the vector-length is calculated at compile-time using PARAMETER statements. This calculation takes into account the total available work-space and the result is also used in setting up the block-gathering and scattering.

Climate modelling has two purposes:-

1. Understanding how the present climate works
2. Assessing the likelihood of climate change due, for instance, to increasing CO_2, a "nuclear winter" etc.

It is necessary to understand how the present climate works, and the relationship between the model and the real climate, before it is wise to use the model to assess how the climate may change. Instantaneous values of variables such as wind, temperature etc are readily available from the model as they are output to disk at fixed intervals of time, but there are many derived quantities that are helpful in diagnosing the behaviour of the model atmosphere which cannot be computed after the model run from these instantaneous values because they depend on the nature of the model evolution between the regular output times. These diagnostics can, however, be computed while the model is running and stored temporarily in memory. At intervals they are converted into the format of the front-end computer and passed through the link for processing and archiving. This system provides a very powerful tool for the research worker in understanding the processes within the model simulation and this will guide him in the evaluation of the model's prediction of climate change. The diagnostic programmes are called from the model as subtasks as are MFLINK and MFQUEUE which pass files and jobs through the link. This increases the elapsed time the model needs to run, but since it greatly increases the understanding of the results, the extra computer time is well worthwhile. A useful pay-off for the run-time processing is that we are able to delete 99% of the output disk-files as we go. As there is a considerable shortage of Cyber disk-space on the Meteorological Office configuration this is very helpful.

When the diagnostic variables were first introduced into the model, the computer only had one megaword of memory, and the storage of the diagnostic variables would have caused further paging. The grid-size was therefore tailored so that the model would fit into the available memory despite the extra diagnostics. Once again PARAMETERs are used at compile-time to allow an appropriate grid-size to be used.

Figure 2 shows the first part of the LOAD map. The large number of routines (which is greater than shown on this diagram) gives an indication of the size and complexity of the climate model programme. The map shows that the data areas of the routines have been kept to a minimum by the use of the shared work-area, the COMMON area IO. The LOAD parameters indicate that the

COMMON areas DCBA, ALBS, IO and BRAD have all been mapped on to large page boundaries to reduce paging.

Figure 3 shows the last part of the LOAD map and the COMMON areas mapped on to large page boundaries appear at the end. Notice the area IO of two large pages, which is used as a work-area by all routines, and also acts as input/output buffers. The COMMON area BRAD, almost a large page in length, starts at address 3800000 which is the start of the 15th large page in memory and therefore it appears that the programme requires 15 large pages of memory. However, because of the manner in which the LOADer works there are two gaps of about three-quarters of a large page (which are marked on Figure 3) and therefore the programme and data will satisfactorily fit into the 14 large pages available to the user.

Part of the dayfile from a model run is given in Figure 4 and shows clearly the sequence of events as the different subtasks are called to process the results. For instance, controllee MN1AO creates the mean of ten daily output files. The model programme tests the return code from controllee MN1AO and if it has been successful the ten individual files are then deleted from disk using Q5PURGE. Controllee MN2AO uses three ten-day mean files to produce a thirty-day mean file, and then the three ten-day mean files can be deleted. In a similar fashion other files can be deleted after a successful MFLINK to the front-end computer.

A later part of the dayfile is in Figure 5 and the time taken for one model-day is shown between 06.18.32 and 06.21.16. Just before this is an MFLINK which took well over a minute to transfer a file. It is likely that this long time (compared with the next MFLINK of 9 seconds) occurred because there was a high priority operational job running on the front-end computer. At 06.22.54 the operator sent the message "STOP" to the model programme because the computer would shortly be needed for operational purposes. At the end of each model-day the programme tests for such a message, and if it finds one it tidies-up and closes down the model run. The tidying-up includes sending three jobs to the front-end computer with MFQUEUES (see dayfile 06.25.28 to 06.26.02). Once of these will resubmit the model to the Cyber 205 and the other two will process the files which have previously been sent through the LINK. These last two jobs will also be submitted to the front-end computer at intervals during long integrations so that there is not too much back-log to be cleared up at one time.

The model includes a TIMER subroutine which is very similar to the FLOWTRACE facility on the CRAY computer. It gives the CPU-time and elapsed time for each routine with a TIMER entry. For other routines the times will be included with the routine at the next higher level. Figure 6 shows typical output from the TIMER routine, and is for the same 15-day integration as the dayfile shown in figure 4 and 5.

It is interesting to note that the routines with noticeable amounts of CPU (DYNEQS, FIFNP, FILTER, HELIOS, BNDARY, RAIN and SHOWER) are all 98% CPU-efficient or better. The "lost" time is all in the Master routine, MODELA01 or in the subroutine FIELDS. The timing for the Master routine also includes that for the output routines for the Cyber-compatible files and the subtasks which make 10-day and 30-day mean files and send them through the link to the front-end machine for archiving. This activitiy is heavily IO-bound. The FIELDS subroutine governs the production of front-end-compatible datasets and sending them through the link for processing and archiving. This is also heavily IO-bound and will include such wasted time as the delay in the link at 0617 already referred to.

The information given so far has all been for the time when the computer had only one megaword of memory. Within a very short time of the commissioning of the second megaword the elapsed time of the model had been reduced by more than 25%, from an average of about 24 hours per model-year to less than 18 hours. There were two primary contributions to this speeding-up, a reduction of CPU-time in the radiation scheme and a reduction in the elapsed time for the subtasks.

The radiation scheme includes a number of routines and the timing is shown under the collective name HELIOS as over 48% of the total CPU-time. This large percentage of the total time is due primarily to the complication of the calculation itself, but is aggravated by the fact that the calculation is still in 64-bit arithmetic and the vector-lengths are very short-typically 72 elements. With the extra memory, the work area IO has been increased in size so that a vector length of 582 is now possible. This saves almost three and a half hours per model-year, which is nearly half the time formerly spent in HELIOS.

When the model itself needed the entire available memory all the time, the calling of a subtask caused considerable paging to occur, with a consequent loss of elapsed time. Although most of the two megawords is now used during the radiation routines, these are only called at intervals of three model-hours, and the subtasks are called at the end of a model-day. Therefore considerably less paging occurs with the extra memory. For instance the 10-day mean subtask, which typically took one and a half minutes (eg 05.53.48 to 05.55.18 in Figure 4) now takes only just over 20 seconds. This reduction in paging saves almost another three hours of elapsed time.

The model is being improved meteorologically all the time, and this frequently means an increase in the CPU-time, but side by side with this, work will continue in an endeavour to reduce the total elapsed time, so that more integrations can be performed in the computer time that is available.

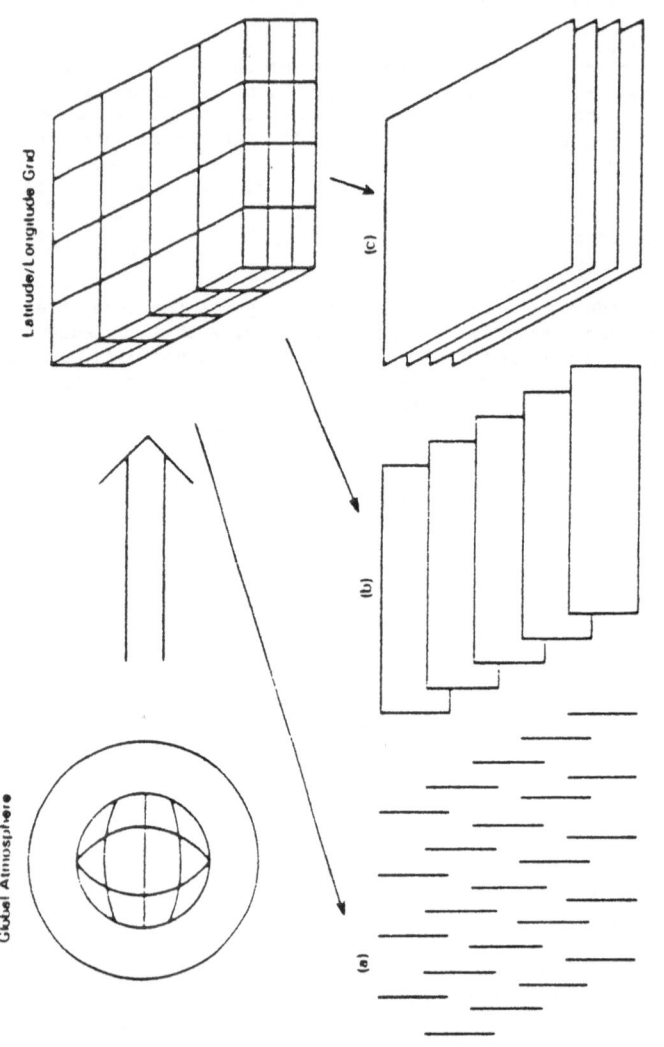

Figure 1 Different ways of organising data for vector computation
(a) vertical columns. (b) vertical slices. (c) horizontal fields

PARAMETERS SPECIFIED: B1,B2,CM=AUCR/1792,CDF=2000,GRLP=DCBA,=ALB5,GRLP==IO,GRSP==IOH,GRLP=BRAD,LIBRARY=DCOBJ.

MODULES

NAME	CODE ADDRESS	WORD LENGTH(HEX)	FILE	DATA BASE ADDRESS	WORD LENGTH (HEX)	PROCESSOR	DATE	TIME
MODELA01	80080	872	B1	A1C80	5F4	FTN2.0	05/1/8	5:28
BRADIMDX	B9A00	42	B1	BAA00	9E	FTN2.0	05/1/8	5:28
BMDIMDX	BD200	4E	B1	BE500	DA	FTN2.0	05/1/8	5:28
AODUMP1O	C1C00	18C	B1	C7E80	1E0	FTN2.0	05/1/8	5:28
AOINITIO	CF700	152	B1	D4800	17C	FTN2.0	05/1/8	5:28
GETHSTAO	DAA80	EA	B1	DE480	CE	FTN2.0	05/1/8	5:28
RDHSTAO	E1880	86	B1	E3980	AE	FTN2.0	05/1/8	5:28
RDVDESAO	E65B0	126	B1	EAE80	7A	FTN2.0	05/1/8	5:28
RDCOMAO	ECD80	86	B1	EEE80	3C	FTN2.0	05/1/8	5:28
RDUPDTAO	EFE00	90	B1	F2180	3E	FTN2.0	05/1/8	5:28
FINDKPSO	F3180	38	B1	F3700	1E	FTN2.0	05/1/8	5:28
FINDJPSO	F4700	38	B1	F5180	1E	FTN2.0	05/1/8	5:28
IBMDSKAO	F5C80	192	B1	FC080	16A	FTN2.0	05/1/8	5:28
CHKLSTAO	101080	32	B1	102780	24	FTN2.0	05/1/8	5:28
MJMDAY	103100	56	B1	104400	36	FTN2.0	05/1/8	5:28
YMDAY	105400	3C	B1	106200	52	FTN2.0	05/1/8	5:28
MMPG1OAO	107780	80	B1	109700	BE	FTN2.0	05/1/8	5:28
MMMMPGAO	10C700	DA	B1	10F600	CE	FTN2.0	05/1/8	5:28
FIELDSAO	113100	296	B1	11B400	232	FTN2.0	05/1/8	5:28
EXECCILE	126300	38C	B1	134500	3C6	FTN2.0	05/1/8	5:28
RAIN	143780	10C	B2	147A00	140	FTN2.0	05/1/8	5:28
FFTG	14CAB0	214	B2	154F00	1B0	FTN2.0	05/1/8	5:28
TIMER	15C380	EA	B2	15FDB0	1DC	FTN2.0	05/1/8	5:28
FIFMP	167500	A2	B2	149000	70	FTN2.0	05/1/8	5:28
DYMEGS	168980	570	B2	181500	6E4	FTN2.0	05/1/8	5:28
FILTER	19CE80	5A	B2	19E480	38	FTN2.0	05/1/8	5:28
POLAR	19F300	58	B2	1A0800	1B6	FTN2.0	05/1/8	5:28
SEIWTS	1A7680	B2	B2	1A2280	98	FTN2.0	05/1/8	5:28
EXP32V	1AC900	46	B2	1ADA00	36	FTN2.0	05/1/8	5:28
BMOARY	1AE800	7C2	B2	1CD900	2FE	FTN2.0	05/1/8	5:28
JRCDATA	1D9800	0	B2	1D97A0	0	FTN2.0	05/1/8	5:28
LAYER	1D9800	E6	B2	1DD100	D42	FTN2.0	05/1/8	5:28
RADIMC	212200	48	B2	213380	46	FTN2.0	05/1/8	5:28
HELIOS2	214580	34	B2	215200	22	FTN2.0	05/1/8	5:28
HELIOS	215800	C8A	B2	247600	C86	FTN2.0	05/1/8	5:28

Figure 2 First part of LOAD map

$Q5CPUTI	685C80	38	56
$Q5ROUTE	68A380	84	180
$Q5RECAL	694E00	3C	60
T	6A0000	0	
HISTORY	6A0000	400	1024
IOH	680000	802	2050
OCEAN	6D0080	76	118
TO	6D1E00	2	2
TRIG	6D1E80	79E	1982
WEIGHTS	6F0E00	FC	252
RADCONST	6F4D00	396	918
RADCHEM	703280	EA	234
NIGHT	706D00	2	2
R	706D80	A380	41904
LOG	995980	804	2052
Q5CMNAL	985A80	12	18
Q5CMNPE	985F00	2	2
Q5_SIOA	985F80	32	50
Q5FIT	986C00	70	112
Q5CMNER	988800	A	10
Q5_CDCDU	988A80	4	4
Q5_CPROE	988B80	3A	58
Q5_EMS01	989A00	2A	42
Q5_EMS04	98A480	E	14
Q5_LFIBF	98A800	4	4
Q5_DCDPF	98A900	1BE	446
Q5_MNE02	9C1880	D0	208
Q5CSNDOP	9C4C80	56	86
Q5ALPHA	9C6200	204	516
Q5_EMS05	9CE300	54	84
Q5_EMS07	9CF800	A	10
Q5CSNDMO	9CFA80	62	98
Q5CGETOP	9D1300	64	100
Q5_EMS02	9D2C00	8	8
Q5_CLFPR	9D2E00	82	130
Q5_MNE01	9D4E80	32	50
Q5_OPTAB	9D5800	DC	220
Q5CTERM	9D9200	34	52
Q5_CLFHI	9D9F00	68	104
Q5CSNDCE	9DB900	58	88
Q5CGETCE	9DCF00	6E	110
Q5CINIT	9DEA80	AE	174
Q5_EMS06	9E1600	64	100
BETA	9E2F00	24	36
Q5CTIME	9E3800	3A	58
Q5CGETCR	9E4680	7E	126
Q5CSNDCR	9E6600	66	102
Q5_EMS03	9E7F80	6	6
Q5_CLFPO	9E8100	72	114
Q5CGETTN	9E9080	46	70 ←gap
99575341	CE0000	2000	8192
	060000	180	384
99464954	D66000	1B80	7040
55000000	DD4000	8010	32784
99415048	FD4400	F78	3960
99434642	1012200	FA	250 ←gap
DCBA	1400000	6D532	447794
ALBS	2F54C80	1B48	6984
IO	3000000	20000	131072
BRAD	3300000	E7E6	59366

ERROR PROCESSING INFORMATION

Figure 3 Last part of LOAD map

```
05.53.46  RHF5401 M20.AUCR154U.PPFIELDS CREATED
05.53.47  ALL DONE
05.53.48  EXECUTE CONTROLLEE MN1A0
05.53.48  (OUTPUT-PRINT,UNIT6-PRINT,UNIT5-MNFILE)
05.53.53  W Q5LFINIR 0304 NO FILES QUALIFY
05.53.59  W Q5LFINIR 0304 NO FILES QUALIFY
05.54.06  W Q5LFINIR 0304 NO FILES QUALIFY
05.54.13  W Q5LFINIR 0304 NO FILES QUALIFY
05.54.21  W Q5LFINIR 0304 NO FILES QUALIFY
05.54.29  W Q5LFINIR 0304 NO FILES QUALIFY
05.54.35  W Q5LFINIR 0304 NO FILES QUALIFY
05.54.44  W Q5LFINIR 0304 NO FILES QUALIFY
05.54.52  W Q5LFINIR 0304 NO FILES QUALIFY
05.54.59  W Q5LFINIR 0304 NO FILES QUALIFY
05.55.15  ** STOP **
05.55.18  ALL DONE
05.55.19  EXECUTE CONTROLLEE MFLINK
05.55.19  (AUCRA54U,ST=PS8,DD=UU,JCS="CREATE,DSN=M20.AUCRA54U,DCB=(BLKSIZE=23476,DSORG=PS,RECFM=VS),DISP=CATLG,UNIT=DISK,VOL=SER=USR4
05.55.21  WAITING FOR CONNECTED STATUS.
05.55.42  RHF5401 M20.AUCRA54U CREATED
05.55.42  ALL DONE
05.55.43  EXECUTE CONTROLLEE ZONN
05.55.43  (OUTPUT-PRINT,UNIT6-PRINT,UNIT5-ZFILE)
05.55.45  W Q5ATTACH 1504 FILE AUCRA54U ALREADY ATTACHED AS A PERMANENT FILE
05.55.46  ZONN UPDATED ON 12 NOV 1982 BY J A BOLTON X2313
05.55.47  ALL DONE
05.55.47  EXECUTE CONTROLLEE FA2PPA0
05.55.47  (AUCRA54U,LIST-PRINT,INPUT-FA2PPIN)
05.55.52  W Q5LFINIR 0304 NO FILES QUALIFY
05.55.55  FIELDS CREATED
05.55.55  ALL DONE
05.55.56  EXECUTE CONTROLLEE MFLINK
05.55.56  (DCCATENP,ST=PS8,DD=US,JCS="CREATE,DSN=M20.AUCRA54U.PPFIELDS,DCB=(RECFM=VBS,LRECL=32000,BLKSIZE=11476,DSORG=PS),DISP=KEEP,'
05.55.57  ONLY 135 CHAR OF JCS TO TRACE
05.55.57  WAITING FOR CONNECTED STATUS.
05.56.10  RHF5401 M20.AUCRA54U.PPFIELDS CREATED
05.56.10  ALL DONE
05.56.11  EXECUTE CONTROLLEE MN2A0
05.56.11  (OUTPUT-PRINT,UNIT6-PRINT,UNIT5-MNFILE)
05.56.15  W Q5LFINIR 0304 NO FILES QUALIFY
05.56.18  W Q5LFINIR 0304 NO FILES QUALIFY
05.56.22  W Q5LFINIR 0304 NO FILES QUALIFY
05.56.32  ** STOP **
05.56.33  ALL DONE
05.56.33  EXECUTE CONTROLLEE MFLINK
```

Figure 4 Part of day file

241

```
06.17.25  EXECUTE CONTROLLEE MFLINK
06.17.25  (DCCATEMP,ST=PS8,DD=US,JCS="CREATE,DSN=M20.AUCR1556.PPFIELDS,DCB=(RECFM=VBS,LRECL=32000,BLKSIZE=11476,DSORG=PS),DISP=KEEP,'
06.17.27  ONLY 135 CHAR OF JCS TO TRACE
06.17.27  WAITING FOR CONNECTED STATUS.
06.18.32  RHFS401 M20.AUCR1556.PPFIELDS CREATED
06.18.32  ALL DONE
06.21.16  EXECUTE CONTROLLEE F42PPAO
06.21.16  (AUCR1557,LIST=PRINT,INPUT=F42PPIN)
06.21.42  FIELDS CREATED
06.21.50  ALL DONE
06.21.51  EXECUTE CONTROLLEE MFLINK
06.21.51  (DCCATEMP,ST=PS8,DD=US,JCS="CREATE,DSN=M20.AUCR1557.PPFIELDS,DCB=(RECFM=VBS,LRECL=32000,BLKSIZE=11476,DSORG=PS),DISP=KEEP,'
06.21.52  ONLY 135 CHAR OF JCS TO TRACE
06.21.53  WAITING FOR CONNECTED STATUS.
06.21.59  RHFS401 M20.AUCR1557.PPFIELDS CREATED
06.22.00  ALL DONE
06.22.54  STOP
06.24.45  EXECUTE CONTROLLEE F42PPAO
06.24.45  (AUCR1558,LIST=PRINT,INPUT=F42PPIN)
06.25.09  FIELDS CREATED
06.25.17  ALL DONE
06.25.18  EXECUTE CONTROLLEE MFLINK
06.25.18  (DCCATEMP,ST=PS8,DD=US,JCS="CREATE,DSN=M20.AUCR1558.PPFIELDS,DCB=(RECFM=VBS,LRECL=32000,BLKSIZE=11476,DSORG=PS),DISP=KEEP,
06.25.20  ONLY 135 CHAR OF JCS TO TRACE
06.25.20  WAITING FOR CONNECTED STATUS.
06.25.26  RHFS401 M20.AUCR1558.PPFIELDS CREATED
06.25.27  ALL DONE
06.25.28  EXECUTE CONTROLLEE MFQUEUE
06.25.28  (JOBFILE,ST=PS8,DD=C8,DC=IX,JCS="SUBMIT")
06.25.33  ALL DONE
06.25.39  W QSCLOSE  1406 FILE TEMPAUCR NOT CURRENTLY OPEN
06.25.39  EXECUTE CONTROLLEE MFQUEUE
06.25.39  (COPYJOB,ST=PS8,DD=C8,DC=IX,JCS="SUBMIT")
06.25.52        1C WORDS OF FILE  COPYJOB  COPIED TO FILE Q5001E4D
06.25.52
06.25.54  ALL DONE
06.25.54  EXECUTE CONTROLLEE MFQUEUE
06.25.54  (COPYJOB,ST=PS8,DD=C8,DC=IX,JCS="SUBMIT"
06.26.00        A7 WORDS OF FILE  COPYJOB  COPIED TO FILE Q5031CF6
06.26.02  ALL DONE
```

Figure 5 Part of day file

FLOW TRACE SUMMARY

	ROUTINE	CPU TIME	% CPU	CALLED	AVERAGE CPUTIME	ELAPSED TIME	% ELAPSED	AVERAGE ELAPSED
1	MODELA01	6.359804	.28	1	6.359804	299.947880	8.99	299.947880
2	READHIST	.023743	.00	1	.023743	.242452	.01	.242452
3	AOINITIO	.013730	.00	1	.013730	4.820995	.14	4.820995
4	DYNEQS	469.713160	20.48	2160	.217460	481.423911	14.43	.222881
5	FIFNP	40.909364	1.78	155520	.000263	41.433537	1.24	.000266
6	POLAR	4.563531	.20	4320	.001056	4.586505	.14	.001062
7	FILTER	75.885419	3.31	77760	.000976	76.812046	2.30	.000988
8	RADINC	1.523109	.07	2160	.000705	2.428449	.07	.001124
9	HELIOS	1109.941471	48.39	120	9.249512	1133.921248	33.99	9.449344
10	BNDARY	156.914277	6.84	6480	.024215	157.461039	4.72	.024300
11	RAIN	119.088087	5.19	6480	.018378	119.185878	3.57	.018393
12	SHOWER	307.675335	13.41	8640	.035611	309.498869	9.28	.035822
13	DIAGTSTR	.582474	.03	2160	.000270	.589546	.02	.000273
14	WRITHIST	.159998	.01	38	.004210	.160124	.00	.004214
15	FIELDS	.462323	.02	17	.027195	703.444553	21.09	41.379091
	*** TOTAL	2293.815825				3335.957032		

Figure 6 TIMER output

EXPERIENCE WITH THE VECTORISATION OF SOME FLUID FLOW PREDICTION CODES ON THE HARWELL CRAY-1

Compiled by

I.P. Jones

Computer Science and Systems Division
AERE Harwell, Oxon OX11 ORA

with contributions from

A.D. Burns, K.A. Cliffe, I.S. Duff, C.P. Jackson, J.R. Kightley,

P.C. Robinson, P.J. Stopford, C.P. Thompson, N.S. Wilkes

SUMMARY

This paper describes the experience obtained at Harwell on the exploitation of the CRAY-1 for the prediction of fluid flows. The topics covered include the use of direct methods and iterative methods for the solution of linear systems of equations, and some other relevant issues such as the use of library kernels and efficient input/output.

1 INTRODUCTION

At the Workshop on the Efficient Use of Vector Processors in Computational Fluid Dynamics there were significant differences in outlook between the participants. Many were trying to obtain the best possible performance from their machine and often struggling to fit the code into the available memory. This is, to a certain extent, in contrast to the type of work performed at Harwell. It is worthwhile therefore explaining some of the background to Harwell so that the reader is able to form a clearer view of the considerations of the author and his colleagues in developing codes for the prediction of fluid flows.

The Atomic Energy Research Establishment at Harwell is Europe's largest centre for multi-disciplinary research and development, and over 60% of its income is currently derived from contract research and development. As a result it has a powerful computer installation with a 2M word CRAY 1S, an IBM 3084Q and several VAX computers. Because of the establishment's background in atomic energy research, fluid dynamics and heat transfer form an important part of its work. This work is derived from both nuclear and non-nuclear applications and covers many fields. For example, the Heat Transfer and Fluid Flow Service (HTFS) is a subscription service for industry and as part of its work it provides codes for the prediction of flows and heat transfer in boilers and furnaces. Other non-nuclear applications have been the development of programs for oil-reservoir simulation, internal combustion engines and dense gas dispersal.

There are several groups active in the development and application of computer codes for flow prediction. These include the Mathematical Modelling Section of the Fluid Phenomena Group in Engineering Sciences Division, the Theory of Fluids Group in Theoretical Physics Division and the Applied Mathematics and Numerical Analysis Groups of Computer Science and Systems Division. The inter-disciplinary nature of the establishment has also encouraged active collaboration and discussion between these groups. This paper describes some of the experience gained by these groups on the effective exploitation of the CRAY-1.

The codes developed at Harwell have used both finite-difference and finite-element methods, and recently there has been some use of a pseudo-spectral code for Large Eddy Simulation. Since this is a recent venture, experience with this approach is limited and will not be discussed here.

The finite-difference work has included both the stream-function vorticity formulation and the velocity-pressure formulation, although attention is now focussed on the velocity-pressure formulation. The applications of this work are now primarily turbulent flows in two and three dimensions using the $k-\varepsilon$ model of turbulence. There is also work being carried out on strongly rotating compressible flow problems. Some of the codes also include models for buoyancy and for combustion.

The finite-element work is based upon the TGSL subroutine library [1] and uses the classical Galerkin formulation.The applications include laminar flows, buoyant flows, simple turbulence models and groundwater flows. These flows may be in either two or three dimensions. The codes have also been extensively used to explore bifurcation phenomena in non-linear flow problems. Examples of the use of these codes may be found in references [2 – 4].

There are many different computers in use within the UKAEA and several of the codes are marketed commercially. For this reason there is a need for these codes to be transportable and efficient across a wide range of computers, both scalar and vector computers. The codes are therefore written in standard Fortran 66 or 77, although a few machine dependent modules, particularly for the CRAY, are permitted. There are, however, standard Fortran versions of these modules. The basic algorithms also need to be efficient on both scalar and vector computers. One bottleneck that has not really been touched upon in this workshop is the human bottleneck. Computer time is comparatively cheap, and, overnight or over a weekend at least, is not usually limited for the kind of problem we wish to tackle. Human effort however is not cheap and is very limited. This effort must therefore be concentrated on those areas of the codes where significant gains may be achieved. Furthermore, it is important not to sacrifice robustness or flexibility for speed of execution since, in the long term, this may be a false economy.

The solution of the flow equations using either finite-difference or finite-element methods can be conveniently divided into several steps:

1. Discretisation and linearisation,

2. Calculation of matrix coefficients and residuals,

3. Solution of linear equations,

4. Non-linear iteration,

5. Interpretation of output.

In many cases the cost of a job is dominated by the cost of the linear algebra. It is an area where a small amount of effort can provide large gains, not just in speed but also in increasing the robustness of the code. Furthermore, by utilising a modular structure for the codes, the routines written for one application may be used in other applications. For this reason a significant effort has been focussed on the solution of linear equations. This paper will therefore concentrate on this topic and will discuss the use of both a direct method, a frontal method, and iterative methods. Some other relevant issues such as input/output and the use of library kernels will also be discussed in the appropriate places.

2 DIRECT METHODS

In many flow problems where Newton's method is used to solve the non-linear systems of equations, up to 90% of the cpu time may go in solving the linearised equations, particularly on large problems with a very wide bandwidth. The matrices that result are also very large and, as a consequence, must be held out of core.

2.1 Frontal method

The frontal method is very convenient and efficient for problems of this type. The basic idea is that the active part of the matrix is held as a small full matrix in core, the frontal matrix, with the remainder being held out of core. The method was developed initially for symmetric equations arising from finite-element applications in structural mechanics. In general, fluid flow problems give rise to unsymmetric matrices and the method was extended by Hood [5] to cope with this situation. This version was significantly improved at Harwell by Andrew Cliffe and Peter Jackson and adopted for use in the finite-element library TGSL [1]. It was later implemented by Iain Duff into the Harwell Subroutine Library under the name MA32. At the same time new facilities were added. These included static condensation and, at the author's request, an entry such that equations could be specified directly, instead of in a finite-element form. This entry means that the subroutine can be used in finite-difference codes.

2.2 Vectorisation of MA32

The inner loop of MA32 is essentially the inner loop of Gaussian elimination for a full matrix and is a direct 'SAXPY'. That is, it has the form $a\mathbf{x} + \mathbf{y}$. Iain Duff noted that the SAXPY routine in the CRAY Scientific Library, SCILIB, had an asymptotic performance of about 44.4 Mflops. A Fortran loop gave an asymptotic performance of about 35.7 Mflops and about 20 Mflops were achieved in practice. However, because of the higher start up time for the SAXPY there was no advantage to be gained from its use on practical vector lengths.

With the help of CRAY Research an assembler coded inner loop which minimised memory references was written. This achieved about 65 Mflops on practical problems. This development is available in the CRAY version of MA32 in the Harwell Subroutine Library. Further details of this code and its vectorisation may be found in Duff [6,7].

The outer loops were now slowing the code down and Peter Jackson suggested combining two elimination steps and chaining the result from one step to the next. This procedure was implemented into an assembler routine by Iain Duff, and, in the light of subsequent practical experience, speeded up further by Peter Jackson and Peter Robinson. The inner loop of their version of MA32 is now achieving speeds of 130 Mflops in practice and, on large problems, average speeds of 100 Mflops are achieved over the assembly and factorisation of the matrix. On these large problems over 90% of the cpu time may be spent in this phase.

2.3 Drawbacks of the frontal approach

The main drawback of this approach is that the problems are restricted to ones where the frontal matrix can be held in core. The 2M words of the current CRAY at Harwell is therefore a little restrictive. Large 2–D problems can be solved with very little difficulty, but in 3–D only small coupled problems, or large ones with only a few variables at each point, can be tackled.

Another drawback of the method is the amount of I/O performed. Typically 50% of the cost of a job which uses MA32 comes from the cpu usage and 50% from the cost of I/O. This is, of course, a function of the charging algorithm. The algorithm in use at Harwell heavily penalises out of core working. Iain Duff has investigated the effect of different forms of I/O on the cost of a simple model problem which was solved using MA32. From this study he concluded that CRAY unblocked I/O was reasonably efficient. and hence it is the method used by MA32 on the CRAY. Full details of this are described by Duff [6].

3 ITERATIVE METHODS

3.1 Linearisation

Iterative methods for the solution of linear equations have been mainly used in finite-difference calculations where the overheads of calculating matrix coefficients and residuals are considerably smaller.

The programs in use at Harwell are based upon the SIMPLE algorithm of Patankar and Spalding, see Patankar [8] for details. With this approach the individual equations are linearised and each is written in the form

$$\nabla.(\mathbf{u} \, \phi) - \nabla.(\, \Gamma \, \nabla\phi) = S \, ,$$

where ϕ is the unknown variable and $\mathbf{u}$ the velocity. Values for the pressure are guessed and the pressure gradient is amalgamated with any other terms representing sources and sinks into the quantity S. The discretised momentum equations are solved for the velocities. The velocity components will not, in general, now satisfy the continuity equation and the pressure and velocity are adjusted using a pressure correction algorithm to ensure mass conservation. Various algorithms have been proposed and the SIMPLE algorithm remains the most popular, although other methods such as the PISO algorithm, Issa [9] are now finding favour. Having satisfied the continuity equation, the equations for the remaining scalar quantities are then solved. This procedure is repeated until convergence of the outer iteration is achieved.

Traditionally these transport equations have been solved using line-relaxation or, in more recent times, using Stone's method, sometimes known as the Strongly Implicit Procedure. Both of these methods contain vector dependencies and therefore, in their standard form, do not vectorise well.

The remaining part of the iterative procedure, the calculation of the coefficients of the linear equations and residual calculations, can be vectorised straightforwardly if the right data structures are used. For example, with a new Harwell code, currently known as FLOW3D, all the loops for the coefficient calculations vectorise. This includes switches between upwind and central differencing, i.e. hybrid differencing, and the decision whether points near to a wall are in the laminar sublayer or in a logarithmic region in the implementation of the law of the wall for turbulent flow. The code also permits arbitrary locations for inlets, outlets, baffles and internal solids. The vectorisation of this was performed by Chris Thompson and Nigel Wilkes and required considerable care in the choice of the data structures. Further details of FLOW3D are given later.

3.2 Pressure correction equation

One of the most important equations in the finite-difference codes is the pressure-correction equation. This ensures that mass is conserved at each iteration. Furthermore, with non-iterative methods such as the PISO algorithm for transient flows, it is important to have accurate solutions of this equation at each time step. For this reason vectorisable methods for the pressure correction equations have been studied in detail.

The pressure correction equations are symmetric and positive semi-definite. They are, in fact, singular because an arbitrary constant can be added to the pressure. These equations are ideal for the use of the Preconditioned Conjugate Gradient method. The standard conjugate gradient method vectorises very well but may have a slow rate of convergence. It is important, therefore, to precondition the system of equations but, unfortunately, recurrence relationships are usually involved in the solution of a preconditioned system of equations. There is therefore a delicate balance between fast convergence and non-vectorisable work. Kightley and Jones [10] have investigated various preconditionings for pressure correction equations arising from 3–D problems. Their conclusions were as follows:

1. There was no best preconditioning to use. The optimal form of preconditioning depended on the particular problem and on the computer.

2. A diagonal scaling followed by classical conjugate gradients vectorised very well and could be good on small problems and on Poisson's equation.

3. The standard Incomplete Cholesky factorisation, ICCG, provided the best compromise overall. It was reasonably good on scalar and vector machines, was reliable on the test problems that were studied and could handle the singularity effectively.

4. A power series expansion due to van der Vorst could give a worthwhile improvement. The ordering of the bands of the matrix became important however and the modification could be significantly worse than ICCG in a few cases.

5. Other modifications, such as imposing constraints on the sum of the coefficients by rows or columns, were affected by the singularity and were not recommended. Nested factorisation, a type of block factorisation, could be good, particularly on a scalar computer but was sensitive to the ordering of the bands.

Full details of this work may be found in Kightley and Jones [10].

3.2.1 Use of CRAY SCILIB

In the course of this work the effects of replacing some of the FORTRAN by 'BLAS' kernels from the CRAY scientific library, SCILIB, were studied. Some of these gave worthwhile savings. For example SAXPY took about 70% of the time of equivalent Fortran code and SDOT about 40% of the time of Fortran. SCOPY, on the other hand, took about 120% of the time of a Fortran routine and its use, at least from the library for COS 1.11, is not recommended. Peter Robinson [11] has carried out an extensive survey of the gains to be achieved by using SCILIB routines. He found that some of the routines were excellent under certain circumstances but that they could be poor in other circumstances. For example, the routine MXMA multiplies two rectangular matrices $m \times n$ and $n \times m$. If $m \gg n$ the performance is excellent but with $n \gg m$ the performance is poor. It is easy, however, to restructure the argument list of the subroutine in the second case and obtain fast performance in both cases.

Preliminary results with the SCILIB library on COS 1.12 indicate that some of the routines have been improved. The reader is therefore advised to carry out his own tests on the appropriate routine in order to establish whether or not it should be used.

3.3 Unsymmetric Transport Equations

The remainder of the discrete transport equations that are solved, those for momentum and turbulence quantities, give rise to unsymmetric coefficient matrices. The simplest method, which also vectorises very well, is the point Jacobi method. Unfortunately this does not converge very quickly and usually the gains from its vector performance do not outweigh its poor convergence.

3.3.1 Solution of tri-diagonal equations

For two-dimensional cases, successive line-relaxation has been very popular, with the sweep taking place in the direction of the flow. This method can easily be extended to three dimensions through the use of a plane relaxation procedure, or an alternating direction procedure. The inner loop of a line relaxation is, of course, the solution of a tri-diagonal system of equations. Two approaches are discussed here, the use of fast scalar assembler code for the solution of the tri-diagonal equation, and the inversion of the order of the loops to put the relaxation process in the inner loop.

3.3.2 Fast assembler code

Jordan [12] has written some assembler coded routines for the solution of tri-diagonal equations, and systems of tri-diagonal equations. The routine TRID is for the solution of a single tridiagonal equation. Jordan showed that this routine was two to three times faster than equivalent Fortran, and, except on problems with very large loop lengths was faster than vectorised code based on cyclic reduction. We have made extensive use of this routine and have obtained worthwhile gains for very little effort. For example for line relaxation on a $31 \times 31 \times 42$ grid, using TRID for the tridiagonal solution, it took 37.3 milliseconds per iteration, with pure Fortran code it took 73.4 milliseconds and with Fortran code and vectorisation inhibited it took 166.4 milliseconds. The inner loop size in this case was 29. These figures show a factor four improvement between the best vector code and code with vectorisation inhibited. This should be compared with other routines in the rest of the code which have no scalar loops and yet, usually, only a factor three to four improvement is obtained between scalar and vector modes on these routines.

3.3.3 Loop inversion

Line relaxation in two-dimensions can also be vectorised in Fortran by changing the nesting of the loops, and then reordering the system of equations to a zebra ordering. Explicit vectorisable calculations may then be carried out with loops sizes half the size of the original loop. Experience with this procedure shows:

1. It is slightly faster per iteration than standard line relaxation using the assembler coded tri-diagonal solver for typical grids with up to say 30 points in each direction. It is not much faster since the loop sizes have been halved in size. One needs to have at least 14 points in the direction of sweep in order to achieve any gains at all.

2. The rate of convergence is often slower than when the conventional direction of sweep is used.

For self adjoint systems of equations with 'property A' the rate of convergence of the equations is asymptotically the same for a conventional sweep or a zebra ordering. In the current case the problem is not self adjoint and there is often a strong flow direction. It is better in this case to sweep in the flow direction and use the assembler tridiagonal solver TRID. Non-asymptotic effects also are important since we are discussing the inner iteration of a non-linear iteration for which high accuracy may not be needed. For these reasons the assembler coded routine is preferred to the use of a zebra ordering.

3.3.4 Other methods

There is a need for better solution techniques for these unsymmetric transport equations, particularly for three-dimensional problems. Stone's method with a fixed iteration parameter often works well since the equations are, in general, strongly diagonally dominant. It does however involve non-vectorisable recurrence relationships and also requires a lot of memory (or out of core working) since there is a need to hold both the matrix and its approximate factors. The non-vectorisable work can be reduced slightly by the adoption of a power series expansion for the factors, see Kightley and Jones [10].

A promising algorithm appears to be the Conjugate Gradient Squared (CGS) algorithm due to Sonneveld, Wesseling and de Zeeuw [13]. This contains little more work than the better known Bi-conjugate Gradient algorithm and should converge up to twice as fast. Preliminary studies by James Kightley using diagonal scaling as a preconditioning indicate that the method has a lot of potential and vectorises well. Per iteration it is about 3 times faster than Stone's method on a $20 \times 20 \times 20$ grid and, for steady-state calculations, only one iteration of either method is required for each outer iteration.

Multigrid methods have also been briefly examined. James Kightley and Chris Thompson [14] have examined the behaviour of two multigrid codes which vectorise well, and also conjugate gradients, on a Poisson equation in 2–D with different grid sizes. The results indicate that if high accuracy for the solution is required then the multigrid methods are better than conjugate gradients. If, on the other hand, only modest accuracy is required then preconditioned conjugate gradients can be competitive because it has a lower start up time. Conjugate gradient methods can also be applied to any combination of grid sizes.

4 THREE-DIMENSIONAL CALCULATIONS OF TURBULENT FLUID FLOW

The code FLOW3D has been written at Harwell to make efficient use of vector and scalar processors for two and three-dimensional laminar and turbulent flow calculations. It provides high level language facilities for problem specification and is transportable over a wide range of computers. The first version to be released is restricted to rectangular or cylindrical geometries and exploits this topology for vectorisation. The calculation of the matrix coefficients is fully vectorisable, including over inlets, outlets, internal baffles and solids. The iterative methods used in the code are those discussed earlier in this paper. This code provides cheap computing of 3–D laminar and turbulent flow problems using realistic grid sizes, and has been used extensively at Harwell on several practical applications. Typical computing times on one ventilation problem were about 2.5 minutes on a $16 \times 16 \times 11$ grid, 15 minutes on a $20 \times 20 \times 27$ grid and about 75 minutes on a $32 \times 32 \times 43$ grid. The $k - \varepsilon$ turbulence model was used in this case so an additional two scalar equations were being solved to determine the eddy viscosity. Further details of the code, and examples of its use, may be found in [15–18].

5 CONCLUDING REMARKS

1. The frontal method is well suited for flow calculations which use Newton's method and vectorises extremely well with an assembler-coded inner loop. However its use is limited to problems where the frontal matrix can be kept in core, and I/O costs, with the Harwell charging algorithm, can account for 50% of the costs.

2. Conjugate gradient methods, particularly for the pressure correction equations, are very useful and give worthwhile gains, both due to their good convergence properties and to vectorisation. From the tests carried out the ICCG variant appears to be a good compromise for the pressure-correction equation.

3. There is also a need to improve the treatment of the coupled system of equations, not just a single equation. Even if the cost of the linear algebra is reduced significantly the calculation of coefficients can take a long time because of the large number of iterations that are taken. Gains obtained in this area will make a substantial difference to costs on both scalar and vector computers.

4. Finally, the author of this note feels that the principal bottleneck to the efficient use of vector computers has not really been discussed in any great detail during this workshop. Computer time is comparatively cheap and human effort of the required calibre is both expensive and limited. The principal costs often arise from the cost of writing, testing, validating and documenting the code, the preparation of data for production runs and the difficulty of interpreting the results for three-dimensional flows. These areas need as much attention as the performance of the codes on vector computers. The principal aim of this work should be, after all, to increase our understanding of the very complex nature of the flows which arise in our individual application areas.

6 REFERENCES

[1] Jackson, C.P., The TGSL finite-element subroutine library, AERE R 10713, 1981.

[2] Winters, K.H., Laminar natural convection in a partially divided rectangular cavity at high Rayleigh numbers, AERE TP 1018, 1985.

[3] Jackson, C.P. and Winters, K.H., A finite-element study of the Benard problem using parameter-stepping and bifurcation search, Int. J. Num. Meth. Fluids, 4 pp127–145 1984.

[4] Cliffe, K.A. and Mullin, T., A numerical and experimental study of anomalous modes in the Taylor experiment, J.Fluid Mech. 153 pp 243–258, 1985.

[5] Hood, P., Frontal solution program for unsymmetric matrices, Int. J. Numer. Meth. Eng. 10 pp 379–399 1976.

[6] Duff, I.S., Enhancements to the MA32 package for solving sparse unsymmetric equations, AERE R 11009 1983.

[7] Duff, I.S., The solution of sparse linear equations on the CRAY–1, AERE CSS 125, 1983.

[8] Patankar. S.V., Numerical heat transfer and fluid flow, Hemisphere, 1980.

[9] Issa, R.I., Solution of the implicitly discretised fluid flow equations by operator-splitting, Imperial College Fluids Section Report FS/82/15 1983.

[10] Kightley, J.R. and Jones, I.P., A comparison of conjugate gradient preconditionings for three-dimensional problems on a CRAY–1, AERE-CSS 162, 1984 To appear Comp. Phys. Comm.

[11] Robinson, P. private communication.

[12]   Jordan, T.L. A guide to parallel computation and some CRAY–1 experiences, in Parallel Computations, edited by G. Rodrigue, pp 1–50, Academic Press, 1982.

[13]   Sonneveld, P., Wesseling, P. and de Zeeuw, P.M., Multigrid and conjugate gradient methods as convergence acceleration, Proc. Bristol Multigrid Conference, 1983, To appear.

[14]   Kightley, J.R. and Thompson, C.P., An evaluation of multigrid software: MGD1V, MGD5V and MG00, in preparation.

[15]   Jones, I.P., Kightley, J.R., Thompson, C.P. and Wilkes, N.S., FLOW3D, a computer code for the prediction of laminar and turbulent flow and heat transfer: Release 1, Proc. HTFS Reserch Symposium, 1985.

[16]   Wilkes, N.S., Thompson. C.P., Kightley, J.R., Jones, I.P. and Burns, A.D., On the numerical solution of 3–D incompressible flow problems, Proc ICFD conference, Reading 1985, also AERE-CSS 179

[17]   Burns, A.D. and Jones, I.P., in preparation.

[18]   Wilkes, N.S. in preparation.

SUMMARY OF THE WORKSHOP

Wolfgang Gentzsch

DFVLR, Institut für Theoretische Strömungsmechanik
Bunsenstraße 10, D-3400 Göttingen, Germany

1. GENERAL REMARKS

During the Workshop two main subjects have been discussed comprehensively: Vectorization and numerical algorithms. Vectorization means adaptation of computer programs and numerical algorithms to the special architecture of vector computers in order to exploit fully their potential, which then often results in remarkable performance improvements.

The aim of this summary, therefore, is twofold. First, we give some statistical data concerning the subjects of the Workshop: Vector machines used, basic equations solved, numerical algorithms implemented and fluid dynamic problems treated. We discuss bottlenecks and summarize the main problems in using vector computers. Second, we extract from the large number of contributions general rules mainly in the areas of vectorization, multitasking, data management and data transfer between core and disk (input/output), portability of the codes and compatibility with scalar machines, which may help in solving other problems more efficiently.

Regarding these subjects, the participants of the Workshop dealt with a variety of problems such as

> vector computer hardware
> vector computer software
> multiprocessors
> multitasking
> compilers
> autovectorizers
> vectorization
> fluid dynamics
> basic equations
> numerical algorithms
> programming aspects
> data management and input/output
> benchmarks and

performance of vector computers.

Therefore this summary may also be regarded as a guide to the several contributions to the Workshop, or as a detailed list of contents.

2. COMPUTATIONAL FLUID DYNAMICS AND SUPERCOMPUTER REQUIREMENTS

During the last two decades computational fluid dynamics (CFD) gained an important position together with experiments in windtunnels and analytical methods. The main objective of CFD is to simulate dynamic flow fields through the numerical solution of the governing equations, e.g. the Navier-Stokes equations, using high-speed computers.

The simulation of two-dimensional (2D) inviscid and viscous flows on vector computers does not represent any difficulties with respect to memory requirements or computation time. In three dimensions (3D), however, one has to compute about 20 to 30 variables per mesh point in a 3D-field per time-step or iteration such as the velocity components, density, pressure, enthalpy, temperature, concentrations, dissipative fluxes, local time steps, geometry coefficients, dummy arrays etc. The computations in the case of 3D are therefore restricted to fairly coarse meshes as well as often not fully converged solutions. The large amount of CPU (Central Processing Unit) processing involved and the fact that the data cannot be held entirely in central memory are the main reasons for the long elapsed times for CFD applications. In these cases, the mapping of the problem onto the architecture of the machine and in particular onto the special organization of the memory must be carefully considered to take full advantage of the vector computer.

The incore possibilities of a vector computer with one million words main memory are fairly restricted, e.g. (depending on the method)

less than $42 \times 20 \times 31$ grid points for Navier-Stokes
less than $20 \times 50 \times 35$ grid points for Euler
less than $161 \times 33 \times 33$ grid points for grid generation.

However, a "Navier-Stokes" mesh has to be very fine, say one million grid points, to understand better the physical flow phenomena, to obtain quantitative results of forces and moments and to study dynamic behavior depending on different parameters. The same is true for the Euler-equations, except that viscous flow is not simulated.

Six examples of very large problems are the Euler-solutions on

- 3×10^5 grid points for the DFVLR F 4 wing [Jain]
- 1×10^6 grid points for the Dillner wing [Rizzi]
- 2.5×10^6 grid points for the Dillner wing [Misegades]

the Euler- and Navier-Stokes solutions on

- 1×10^5 grid points for hemisphere cylinder [Barton]

and the Navier-Stokes solutions on

- 0.9×10^5 grid points for hemisphere cylinder [Kordulla]
- 2.1×10^5 grid points for hemisphere cylinder [Kordulla].

Some authors [e.g. Barton, Kordulla, Misegades, Rizzi] presented several dense-mesh solutions with much richer details e.g. in the vortex-shock-wave structure or the separation region for the fine mesh solutions.

The largest problem with 2.5×10^6 grid points has been performed on a CRAY X-MP/48 vector computer with a 128 million word Solid-state Storage Device (SSD). This mesh required approximately 7 million words of internal memory and 100 million words of external memory. The CPU time for convergence at 800 iterations was 7.9 hours, elapsed time was 8.3 hours [see Misegades].

The 1×10^6 grid problem has been performed on a CYBER 205 with 16 million 64-bit words main memory [Rizzi]. The data set was 23 M 32-bit words. The CPU time here for 1000 iterations (the computer codes used by Misegades and Rizzi are somewhat similar) was 2.5 hours with 2.5 hours elapsed time.

From these examples it is obvious, that only the fastest supercomputers, namely vector and parallel computers, enable us to solve such large problems in a reasonable time. It is therefore most important to nearly optimally adapt codes and algorithms to the vector or parallel computer in use. One conclusion is [Schönauer], that we need well trained users, much better trained than for (scalar) general purpose computers.

3. VECTOR COMPUTERS

We distinguish between two principal architectures presently in use, namely register-to-register machines and memory-to-memory machines [Schönauer]. The vector computers used by the participants of the Workshop were

- CRAY-1M/1S [Hemmerich, Jain, Jones, Katzer, Kordulla, Lecointe, Montagne, Volkert, Schönauer]

- CRAY X-MP/11/22/48 [Barton, Dave, Jain, Löhner, Misegades, Schönauer]

- CYBER 205 with 1, 2 and 4 pipes [Hinds, Koppenol, Löhner, Rizzi, Schönauer, Wubs]

- FUJITSU VP 200 [Jain]

Here, the CRAY and FUJITSU machines are register-to-register machines and the CYBER 205 memory-to-memory machines. A brief description of the hardware of these vector computers has been given in [Dowling, Jain, Misegades, Schönauer].

Furthermore, CPU processing times and rates of data processing (RDP) per time step per grid point (or volume) for the above mentioned computers and several numerical algorithms may be found in [Jain, Katzer, Koppenol, Kordulla, Lecointe, Löhner, Montagne, Rizzi, Schönauer, Volkert], while Hockney's $n_{1/2}$, the theoretical half performance length (i.e. vector length for which the vector processor reaches half of its maximum speed), has been discussed in [Dowling, Schönauer, Wubs].

4. MAIN BOTTLENECKS OF THE VECTOR COMPUTERS

Severe bottlenecks were only discussed for the CRAY-1M/1S and for the CYBER 205. However, the problem of excessive input/output transfer between CPU and main memory, above all for 3D solutions, has been pointed out by nearly all the participants. The main bottlenecks for the CRAY-1M/1S are:

- only one memory access port from main memory to vector registers which results in one word transfer per cycle

- memory bank conflicts (for noncontiguous vectors of constant stride)

256

- no gather/scatter hardware instructions

- excessive I/O for out-of-core programs

- poor autovectorizer (up to CFT 1.11)

- FORTRAN 77 not very suitable for vector operations

for the CYBER 205:

- vector length 64 K-1 = 65535

- vector contiguous in memory

- long start-up for short vectors

- no fast secondary storage

- excessive paging for large problems

- poor autovectorizer

- FORTRAN 77 not very suitable for vector operations .

Many of the above mentioned bottlenecks have been overcome for the CRAY X-MP and the VP 200 [Dave, Jain, Misegades, Schönauer], and some for the CYBER 205 [Hinds, Löhner, Rizzi]. However, no autovectorizer or preprocessor can produce from a "bad" program a "good" vector program. The user must devise good programs himself [Schönauer].

5. BASIC EQUATIONS

The main problem in CFD is to solve the basic system of nonlinear differential equations, called the Navier-Stokes equations - consisting of the continuity, momentum and energy equations - by numerical methods on high speed computers. Because of the complicated nature of these equations, simpler models have also been derived and solved numerically.

While the full potential equation models inviscid irrotational flows, the Euler equations describe inviscid flows with rotation. On the other hand, viscous incompressible and compressible, laminar and turbulent flows with e.g. separation are best characterized by the Navier-Stokes equations. In the following, we give a list of the different governing equations solved by the participants:

- 2D/3D Poisson equation for different purposes (grid generation, pressure iteration, Stokes problem, [Hemmerich, Jain, Lecointe, Volkert])

- 3D full potential equation [Jain]

- 2D Euler equations [Löhner, Montagne]

- 3D Euler equations [Barton, Jain, Koppenol, Löhner, Misegades, Montagne, Rizzi]

- 2D Navier-Stokes equations in primitive variable formulation [Katzer]

- 2D Navier-Stokes equations in velocity-vorticity formulation [Schönauer]

- 2D Navier-Stokes equations in vorticity-stream-function formulation [Lecointe]

- 2D shallow-water equations [Wubs]

- 3D Navier-Stokes equations in thin-layer formulation [Barton, Kordulla]

- 3D Navier-Stokes equations (Stokes-problem with incompressible convection equation), [Hemmerich]

- 3D Navier-Stokes equations in velocity-vorticity formulation [Schönauer]

- 3D Navier-Stokes equations in primitive variable formulation [Kordulla, Volkert].

6. BOTTLENECKS CAUSED BY THE EQUATIONS AND THE PHYSICS

Because of the nonlinearity of the equations, the boundary conditions, the discontinuities, the complex geometries, the need for time-accurate results for unsteady phenomena, etc. the numerical solution on vector computers becomes very difficult. Some important drawbacks of the special equations for the full potential equation are:

- type-dependent discretization for subsonic/supersonic mesh points

for the Euler equations:

- time-dependent calculation for the steady state of transonic flows resulting in long-time computations

- non-physical (extrapolation) boundary conditions at walls, outflow and farfield

- need for artificial viscosity or one-sided differences (of lower order)

- strong gradients near shocks

for the Navier-Stokes equations:

- involve second order terms

- thin boundary layer for high Reynolds number flows

- need for high resolutions in regions with strong gradients, separation, transition, etc.

- turbulence modelling

- partly non-physical (extrapolation) boundary conditions

- div $U = 0$ for incompressible flows and pressure correction.

These drawbacks have been discussed and, to some extent, overcome by the authors mentioned in the previous section.

7. NUMERICAL ALGORITHMS

One possible classification of vector algorithms has been proposed in [Schö-nauer] dividing the algorithms into the following three subgroups:

- unchanged general purpose computer algorithms with many recursions, short vector length, which have not been designed for vectorization

- "vectorized" general purpose computer algorithms which solve the 2D or explicit 3D problems very fast, but are not suited for large data sets

- "vectorized" algorithms for large implicit 3D problems, with optimal data structure for out-of-core and performance not limited by the speed of the vector pipes but by the I/O bottleneck (see section 11) of most of the vector computers.

Unfortunately, the old general purpose computer software (first group) wastes the power of the vector computers which grows considerably with the skill of the user [Schönauer].

Various numerical algorithms belonging to these groups have been implemented on vector computers. A detailed description of the vectorization of most of the following algorithms may be found in the individual contributions of this book. We therefore restrict ourselves to list the corresponding algorithms:

- Conjugate gradient methods and variants, e.g. ICCG, BI-CG [Hemmerich, Jones, Schönauer]

- FIDISOL - a finite difference solver [Schönauer]

- successive line overrelaxation [Jain, Jones]

- LU - decomposition [Jain, Schönauer]

- LL^T - decomposition [Hemmerich]

- finite volume spatial discretization [Jain, Katzer, Kordulla, Rizzi]

- Runge-Kutta time stepping [Hemmerich, Jain, Rizzi, Wubs]

- MacCormack schemes [Katzer, Kordulla, Montagne]

- van Leer scheme [Montagne]

- Approximate tri-diagonal factorization of Beam and Warming [Barton, Lecointe] and bi-diagonal factorization of MacCormack [Kordulla]

- FLO 22 [Jain]

- FLO 57 [Misegades]

- Osher's Riemann solver [Koppenol]

- finite elements [Hemmerich, Löhner]

- fourth order space approximation [Lecointe, Wubs]

- alternating direction implicit [Lecointe, Wubs]

- operator compact implicit [Lecointe]

- Thomas algorithm [Lecointe]

- Uzawa algorithm for Stokes problem [Hemmerich]

- Chorin's method for pressure correction [Hemmerich]

- Frontal method for linear systems [Jones]

- TEACH code [Jones]

- Fast Fourier transformation [Volkert]

- Poisson solver POISSX, POISSV [Volkert].

A strategy of vectorization together with some important suggestions how to overcome the main drawbacks caused by the algorithms will be given in section 9.

8. FLUID DYNAMIC PROBLEMS ON VECTOR COMPUTERS

A last interesting topic of these statistical data are the fluid dynamic problems treated by the participants on the different vector computers. Because not all fluid dynamic applications may be recognized in reading the titles of the contributions, we give here a complete list:

- 2D heat driven cavity [Schönauer]

- 3D convective flow [Schönauer]

- 3D full potential code for transonic flow past wings [Jain]

- 3D grid generation method [Jain]

- Transonic flow past ONERA M6 wing using Euler equations [Jain]

- Transonic flow past Dillner wing [Misegades]

- Leading-edge vortex flow around Dillner wing [Rizzi]

- Flow past airfoils [Lecointe, Montagne]

- 2D shallow water flow problems [Wubs]

- 2D shock boundary layer interaction [Katzer]

- 2D flow past a cylinder [Lecointe]

- Transonic/supersonic channel flow [Koppenol, Löhner]

- 3D transonic flow past a hemisphere-cylinder [Barton, Kordulla]

- 3D flow around cylinder [Hemmerich]

- Cooling tower outfall problem [Hemmerich]

- Laser fusion problem [Dave]

- Taylor-green vortex [Volkert]

- Atmospheric flows with long time scales (climate models) [Hinds].

9. VECTORIZATION

The reduction of the CPU processing time may be achieved by vectorization,
i.e. by adaptation of the computer programs or numerical algorithms to the
special vector architecture, and, if possible, by multitasking the computer

program, i.e. by splitting the program into distinct tasks to run on two, or more, processors simultaneously (see section 10).

Because of the deficiencies of the software available on vector computers, above all the limited capabilities of the autovectorizing compilers [Jain, Montagne, Schönauer], the CFD user has to vectorize his fluid dynamic code. With the following strategy, an efficient vectorization of the most time-consuming parts of a computer program may be achieved on today's vector computers within an acceptable time:

Step 1:
Generation of a histogram showing the amount of CPU processing time for different sections of the program. Furthermore, at the end of each subroutine, the compiler presents a complete list of the vectorized and non-vectorizable DO loops together with the reasons for non-vectorizability.

Step 2:
Hand-tailoring of the most time-consuming subroutines if autovectorization does not suffice. It should be mentioned here that a highly modular code obtained by structural programming and consisting of short subroutine modules may, in general, be vectorized more easily [Katzer].

Step 3:
In the case of highly serial algorithms involving, for example, linear or nonlinear recurrences, step 2 is not successful, and a restructuring or even a complete replacement of the algorithm is necessary [Volkert]. In some cases, ASSEMBLER coding of parts of the program may be recommended [Dowling and Katzer].

A "checklist" for efficient vectorization especially at DO loop level (step 2) contains, among others, *depending on the computer*, the following items:

• make:
Innermost DO-loops most efficient

• avoid:
IF-statements
Subroutine and function calls
Irregular (nonlinear) addressing
Memory bank conflicts
Excessive paging
Complicated branching within a loop
Linear and nonlinear recursions

• use:

Many arithmetic operations within inner loops instead of many loops with few operations
Long vectors (i.e. large one-dimensional arrays with length equal to the number of grid points, instead of 2D arrays)
Unformatted i/o
Chaining possibilities and linked triads to achieve supervector speed (done by some compilers automatically)

• unroll:
short DO-loops

• separate:
Vectorizable from non-vectorizable parts

• switch:
Inner and outer loops to have the longer index range on the inner loop and to suppress dependencies .

Some important points concerning step 3, vectorization of numerical algorithms (see section 7), with regard to vector computers that have been discussed during the Workshop are

• For the CYBER 205, change 2D and 3D arrays to 1D arrays [Katzer, Koppenol]

• Use diagonal storing for banded matrices [Schönauer]

• Avoid gather/scatter on the CRAY-1S [Hemmerich]

• Preconditioning in ICCG with approached inverse [Jones]

• Assembler programming on the CRAY-1S of the most time-consuming parts [Dowling]

• Use of CAL - or Q8 - routines for CRAY resp. CYBER [Hinds, Jain, Jones, Katzer, Löhner, Montagne]

• Care has to be taken in implementing the boundary conditions on vector computers to maintain long vectors [Katzer, Rizzi]

• Solve many algebraic systems simultaneously (as in SLOR, ADI, POISSV) [Jain, Lecointe, Volkert]

- Simplify algorithms using irregular addressing (such as pivoting in Gaussian elimination)

- Use red-black or zebra reordering in point or line relaxations [Dave, Jones]

- Use simple elements (triangles in 2D, tetrahedra in 3D) and, if possible, hardware gather/scatter in finite element codes [Löhner]

- Separate the computations of subsonic and supersonic points in transonic calculations [Jain]

- For unstructured grids use explicit schemes [Löhner]

- Improve the stability of explicit schemes which are very suitable for vector and parallel computers.

Some further notes concerning the last point about explicit methods should be made. Explicit schemes have some remarkable properties [Hemmerich, Jain, Katzer, Löhner, Misegades, Rizzi, Wubs]. Besides the better representation of the physical behavior of waves with finite speed, they imply low storage, easy programming and excellent time-accuracy for transient problems. Moreover, they have the advantage of being fully vectorizable, because there are no recursive solution processes and the length of the vectors can be made as long as the number of computational points or cells [Rizzi]. They are easy to divide into subdomains and, therefore, easy to multitask.

A disadvantage for solutions which vary slowly in time and for steady state calculations, however, is the restriction of the time step due to restrictive stability requirements, which results in a low convergence rate. It is, therefore, necessary to relax this restriction and improve the convergence of explicit schemes by, e.g.

- Optimal Runge-Kutta stability bounds

- Enthalpy damping

- Local time-stepping and domain-splitting techniques which allow advancing the solution with different time steps at different grid points, cells or regions of the mesh

- Higher order spatial discretization, in which case the number of grid points may be reduced (if the solution is smooth enough) and thereby the time step increased

- Superstep acceleration techniques for parabolic type problems, the time steps being calculated from the reciprocals of the zeros of certain damped Chebyshev polynomials

- Explicit or implicit averaging of the residuals

- Multigrid or simple multiple grid techniques.

10. MULTITASKING

Whilst vector processing takes place at the level of innermost DO loops, the division of a job into sub-tasks for multitasking should be as global as possible. Three strategies may be applied when simultaneously using two or more processors [Barton, Dave]:

- Domain decomposition: Divide the computational domain into substructures to obtain independent problems in each subdomain to be solved on different processors.

- Program decomposition: Split the computer code into smaller parts (e.g. subroutines with similar amount of CPU processing time) each of which is then solved on one processor.

- DO-loop decomposition: same as program decomposition, but on DO-loop level.

One efficient tool is dynamic load balancing [Dave] to distribute the work equally among the different processors. Before multitasking, however, it is recommended to vectorize the computer program for the single processors.

For some special CFD applications, the concept of concurrently using multiple CPU's for a single program has led to a speedup factor of up to 3.5 on a 4-processor computer, as compared to the vectorized, non-multitasked codes. A detailed analysis of speed-up is discussed in [Barton] and compared with the corresponding actual speed-up for real computations.

11. INPUT/OUTPUT TRANSFER

Another problem discussed in nearly all contributions to the Workshop is the excessive input/output transfer between CPU and (secondary) out-of-core storage, which is due to the limited amount of central memory and/or the extreme data size arising mainly in the computation of 3D problems.

3D Euler solutions, for example, on meshes containing about 50 000 grid points [Jain] and 3D Navier-Stokes solutions on meshes containing about 20 000 grid points [Kordulla], can be obtained on a machine with one million words of main memory without using extended storage on disk. Further mesh refinement, however, is necessary in order to validate the flow model and to better understand the physics of the resulting numerical solutions [Kordulla, Rizzi]. Therefore, for almost any 3D application, the amount of data to be handled exceeds the core memory size except on large memory machines. Only the active data can be kept in central memory while all the remaining data have to be stored on out-of-core storage devices, e.g. on discs. This results in rather long i/o waiting times. In order to avoid, to reduce or to speed up i/o transfer for these data during the computations, the following proposals have been made by the participants of the Workshop [see e.g. Hinds, Jain, Kordulla, Löhner, Misegades, Montagne, Rizzi, Schönauer, Volkert]:

- Split the computational grid into a few blocks of grid points, often called zones, substructures, subdomains or simply "blocks". All variables are then stored in arrays with length equal to the number of points in one zone. Two examples are the pencil concept and the plane concept. The pencil concept is used to arrive at long vector strings. In the plane concept a certain number of planes are contained in the core of the computer, such that all terms needed in the governing equations can readily be evaluated.

- Separate the optimal data selection from the processing of the data, and develop a data flow algorithm to optimally handle data blocks in actual use for the program, e.g. by dynamic data management routines. Strategies used to achieve efficient i/o operations in this context are random direct access file, unblocked buffered i/o, file banking, i.e. parallel i/o with several disk channels, [e.g. Montagne, Schönauer].

- Use of several time step calculations in each block before changing to the next block [Jain].

- In some cases, i/o may even be avoided by recomputing some quantities at each iteration (e.g. quantities depending only on the mesh and not on the solution field) rather than to read them from an auxiliary file after an initial computation [Rizzi]. However, the storage of the solution, and the necessary number of iterations or time levels, depending on the scheme used, limit the possible reduction of the i/o.

- Perform 32-bit words calculations on CYBER 205 as often as possible [Rizzi, Wubs] and 64-bit words only where necessary. This not only results in a further reduction of i/o but also speeds up the CPU processing time.

- Use faster out-of-core storage devices or vector machines with a much larger main memory. The CRAY Solid-state Storage Device, for example, transfers data at rates up to 100 times faster than the fastest disk drives. The use of the 128 million word SSD for a problem with 2.5 million grid points reduced the i/o transfer waiting time remarkably [Misegades] Another example, for in-core calculation [Rizzi], presented at the Workshop was a CFD problem with one million grid points solved entirely in the core of a CYBER 205 with 16 million 64-bit words (or 32 million 32-bit words) of main memory.

12. PORTABILITY

One last point discussed at the Workshop is the difficult problem of generating portable software to maintain one code for architecturally different computers without loss of efficiency. This is not yet possible with Fortran 77, thus special versions for different computers must be generated [Schönauer].

A portable source code should contain only explicitly vectorized parts being performable by the auto-vectorizers. Machine-specific properties of the code should be concentrated into only a few subroutines. By selecting a particular parameter in the compilation run, one can activate either code (subroutines) for a scalar machine or for a special vector machine [Koppenol]. Besides, it is often seen that the vectorizable code is more efficient even in scalar mode or on a scalar machine.

In most cases, however, with today's supercomputers, efficiency predominates over the problem of portability of scientific codes.

13. NEAR FUTURE

Besides improvements in

- CPU processing time and

- much larger central memories

for the next generation of vector computers such as the CRAY-2 with 4ns cycle
time, 4 processors, up to 256 MWords central memory and up to 2 GIGAFLOPS, or
the CDC ETA[10] with 5ns cycle time, up to 8 processors and up to 6.4 GIGAFLOPS,
the computational fluid dynamicists of the workshop expect some further improvement compared to the present situation of vector computer hardware and
software. The main progress will be expected in

- FORTRAN 8X, which should be much more suitable for vector computers
  [Schönauer]

- Autovectorizers, which should vectorize many of the items mentioned in
  the "checklist" of section 9

- "Automultitaskers", which automatically distribute the work equally among
  the different processors

- Capability of managing problems with complex geometries [Hemmerich] or
  even unstructured grids [Kordulla, Löhner] which implies handling of irregular data structures

- More detailed diagnostics to better support vectorization, multitasking
  and I/O .

On the other hand, mathematicians, physicists and engineers have to successively improve and develop numerical algorithms which

- are more accurate, therefore getting along with less grid points and less
  storage requirements

- are better suited for vector and parallel processors and

- have a much faster convergence rate than today's algorithms.

Finally, a very close cooperation of the CFD users and the advisers of the
vector computer manufacturers is of growing importance.

List of Participants

Dr. John T. Barton, NASA Ames, NAS Project Office, Mail Stop 233-1, Moffet Field, California, 94035, USA.

Dr. Ameet Davé, CRAY Res. (U.K.), London Road, Bracknell, Berks., RG12 2SY, Great Britain.

Michael Dowling, TU Braunschweig, Inst.f.Angew.Math., Abteilg. Rechentechnik, Pockelstr. 14, D-3300 Braunschweig, West-Germany.

Dr. Wolfgang Gentzsch, DFVLR SM-TS, Bunsenstr. 10, D-3400 Göttingen, West-Germany

Ingenieur Philippe Hemmerich, EDF, Dir. des Et. et Rech., 1 Avenue du Général de Gaulle, F-92141 Clamart CEDEX, France.

Miss Mavis K. Hinds, U.K. Meteorological Office, London Road, Bracknell, Berkshire, RG12 2SZ, Great Britain.

R.K. Jain, DFVLR, Inst.für Entwurfsaerodynamik, Postfach 3267, D-3300 Braunschweig-Flughafen, West-Germany.

Dr. Ian P. Jones, AERE Harwell, Building 89, Computer Science Div., Didcot, Oxon OX11 ORA, Great Britain.

Dipl.-Math. E. Katzer, DFVLR WB-AE, Bunsenstr. 10, D-3400 Göttingen, West-Germany.

Ir. P.J. Koppenol, National Aerospace Laboratory NLR, Anthony Fokkerweg 2, NL-1059 CM Amsterdam, The Netherlands.

Dr. Wilhelm Kordulla, DFVLR SM-TS, Bunsenstr. 10, D-3400 Göttingen, West-Germany.

N. Kroll, DFVLR, Inst.für Entwurfsaerodynamik, Postfach 3267, D-3300 Braunschweig-Flughafen, West-Germany.

Dr. Yves Lecointe, Groupe Modelisation Numerique, Laboratoire D'Hydrodynamique Navale, ENSM, 1 Rue de la Noe, F-44072 Nantes CEDEX, France.

Rainald Löhner, Dept. of Civil Eng., University College of Swansea, Singleton Park, Swansea SA2 8PP, Great Britain.

Kent Misegades, CRAY Res., Applications Dept., 1440 Northland Drive, Mendota Heights, MN 55120, USA.

Jean Louis Montagné, ONERA, 29 Avenue de la Division Leclerc, F-92320 Chatillon, France.

Prof. Dr. Arthur Rizzi, FFA, Box 11 021, S-16111 Bromma, Sweden.

Dipl.-Math. Eric Schnepf, Rechenzentrum der Universität
    Karlsruhe, Postfach 6380, D-7500 Karlsruhe 1, West-
    Germany.

Prof. Dr. Willi Schönauer, Rechenzentrum der Universität
    Karlsruhe, Postfach 6380, D-7500 Karlsruhe 1, West-
    Germany.

J.P. Therre, Swiss Federal Institute of Technology, Dep. of
    Mech. Eng., CH-1015 Lausanne, Switzerland.

Dr. Hans Volkert, DFVLR, Inst.für Physik der Atmosphäre,
    Oberpfaffenhofen, D-8031 Wesseling/Obb., West-Germany.

Ir. Friederik W. Wubs, Centre for Mathematics and Computer
    Science, Kruislaan 413, NL-1098 SJ Amsterdam, The
    Netherlands.